W9-CEP-062

THE ELEMENTS OF COUNSELING

Seventh Edition

THE ELEMENTS OF COUNSELING

SEVENTH EDITION

Scott T. Meier
University at Buffalo

Susan R. Davis
Private Practice, Buffalo, NY

BROOKS/COLE
CENGAGE Learning

Australia • Brazil • Japan • Korea • Mexico • Singapore • Spain • United Kingdom • United States

BROOKS/COLE
CENGAGE Learning™

The Elements of Counseling, Seventh Edition
By Scott T. Meier and Susan R. Davis

Acquisitions Editor: Seth Dobrin

Editorial Assistant:
Rachel McDonald

Assistant Editor: Nicolas Albert

Senior Marketing Manager:
Trent Whatcott

Marketing Coordinator:
Darlene Macanan

Senior Marketing
Communications Manager:
Tami Strang

Editorial Production Manager:
Matt Ballantyne

Creative Director: Rob Hugel

Senior Art Director:
Caryl Gorska

Manufacturing Director:
Barbara Britton

Manufacturing Buyer:
Paula Vang

Permissions Editor, Image:
Leitha Etheridge-Sims

Permissions Editor, Text:
Bob Kauser

Production Service:
PrePress PMG

Cover Designer: Jeremy Mende

Cover Printer: Thomson West

Compositor: PrePress PMG

©2011, 2008, 2005 Brooks/Cole, a part of Cengage Learning.

ALL RIGHTS RESERVED. No part of this work covered by the copyright hereon may be reproduced or used in any form or by any means—graphic, electronic, or mechanical, including photocopying, recording, taping, web distribution, information storage and retrieval systems, or in any other manner—without the written permission of the publisher.

For product information and
technology assistance, contact us at **Cengage Learning
Customer & Sales Support, 1-800-354-9706.**
For permission to use material from this text or product,
submit all requests online at **www.cengage.com/permissions.**
Further permissions questions can be e-mailed to
permissionrequest@cengage.com

Library of Congress Control Number: 2009934444

Student Edition:
ISBN-13: 978-0-495-81333-0

ISBN-10: 0-495-81333-8

Brooks/Cole
20 Davis Drive
Belmont, CA 94002-3098
USA

Cengage Learning is a leading provider of customized learning solutions with office locations around the globe, including Singapore, the United Kingdom, Australia, Mexico, Brazil, and Japan. Locate your local office at **www.cengage.com/international.**

Cengage Learning products are represented in Canada by Nelson Education, Ltd.

To learn more about Brooks/Cole, visit **www.cengage.com/brookscole.**

Purchase any of our products at your local college store or at our preferred online store **www.ichapters.com.**

Printed in the United States of America
1 2 3 4 5 6 7 13 12 11 10 09

*In memory of Oliver Meier, 1928–1995,
and Donald Davis, 1928–2000*

CONTENTS

CHAPTER 5
Counselor, Know Thyself   56

CHAPTER 6
A Brief Introduction to Intervention   63

PREFACE

Strunk and White's (2000) *The Elements of Style,* a basic introduction to English composition, provided the model for this book. As White notes in his introduction, Strunk originally composed *The Elements of Style* to "cut the vast tangle of English rhetoric down to size and write its rules and principles on the head of a pin" (p. xiii). Strunk, a university English instructor, attempted to produce a set of rules to help students avoid basic mistakes in composition. Our purpose is similar: to distill the basic elements of counseling and teach what counseling is as well as what it is not.

The *Elements of Counseling* aims to present information that is essential for beginning counselors to know and for experienced counselors to remember. Many instructors use *Elements* as an advanced organizer for subsequent instruction and practice, to provide a way to think about counseling, and to clarify the nature of the counseling process. Consequently, *Elements* should be useful for graduate students in the helping professions (such as psychiatry, psychology, social work, and counseling), as well as for paraprofessionals and peer counselors who are learning basic counseling, communications, and listening skills. Undergraduate students enrolled in introductory psychology or counseling courses will find the book useful in helping them understand the applied aspects of psychology. And given the importance of listening and communication skills in their professions, the book will also be of use to nurses, physicians, educators, police officers, and business persons.

In addition to beginners, we also hope that advanced students and practicing therapists find these ideas to be useful reminders. Some practicing counselors, coping with a significant workload and lacking regular supervision, can sometimes lose a degree of professional discipline. In such circumstances one can easily fall into ineffective roles (for example, become maternal/paternal with clients) and forget that progress in counseling can depend as much on what the counselor does *not* do as well as what she or he does. In addition, many of this book's concepts can be considered intermediate outcomes, that is, necessary steps in a sequence of events on the way to a successful resolution of the client's major problems. Factors such as a poor working alliance, insufficient self-disclosure by the client, and inappropriate therapist demands have all been associated with treatment failure (Mash & Hunsley, 1993). Such topics form the gist of this book and may be key elements to attend to when clients fail to progress.

Describing counseling's basic elements remains a difficult task. Counseling lacks a strong consensus about such fundamental issues as the integration of diverse counseling approaches (Ivey, 1980), the usefulness of counseling research to practice (Gelso, 1979; Goldman, 1976; Rice, 1997; Trierweiler & Stricker, 1998), and the interaction between various client characteristics and counseling approaches (Krumboltz, 1966). More than 400 types of psychotherapy are said to be available (Kazdin, 1994; Perez, 1999) and over 550 treatments employed with children and adolescents (Kazdin, 2000). This lack of consensus is understandable given the relative youth of the helping professions and the complexity of the undertaking.

Moreover, this disunity is shared by most other branches of psychology and the social sciences (see Meier, 1987), and increased attention is being focused on such issues (Prochaska, Johnson, & Lee, 1998; Wachtel & Messer, 1997; Zeig, 1997). It is important to note that the profession does possess a knowledge base and that a consensus exists that counseling works. However, theory, research, and practice in counseling often appear fragmented and contradictory.

If such disunity in counseling exists, what is there to teach in a basic elements book? Our sense is that much of the disagreement centers on counseling interventions; counselors share greater agreement about the initial stages of counseling. For example, most counselors would agree that information must be gathered about the client and that some degree of rapport must be established between client and counselor. Building on the work of others, this book represents our attempt to delineate these basic elements of counseling practice. Consequently, *The Elements of Counseling* focuses on listening skills, relationship building, counseling process, and self-exploration, the foundations upon which further intervention is laid.

Although some agreement exists among counselors about initial counseling practices, consensus has yet to be reached about the timing, sequence, or relative importance of these practices (Goldfried, 1983; see also Cormier & Cormier, 1998). Thus, different counselors and counseling instructors might rearrange substantial portions of this book to suit their preferences and experiences. For example, counselor educators who hold counselor characteristics and the counselor-client relationship as central might consider assigning "Counselor, Know Thyself" as the first chapter to be read. This book could be read in sections or in sequence, as with a textbook.

Given the book's niche as a brief primer and the continued positive response to the first six editions, we minimized changes that would add substantially to the book's length. We revised and added material in the section on culturally competent counselors, noting the increased emphasis in the counseling professions on the effects of globalization and the need for counselors to be aware of military culture as they work with veterans and their families. Chapter 6 has a new section on "New and Emerging Approaches" to counseling that briefly describes several newer methods that are innovative in their approach to intervention. And we also added a new section in Chapter 4 that presents an overview of an important issue for many clients, that of grief, loss, and trauma.

Although *The Elements of Counseling* presents guidelines and rules, one cannot become a counselor simply by memorizing them. Given the necessarily brief nature of *The Elements of Counseling,* students may overestimate or underestimate their capabilities if they read only this book. Practice in role plays, actual experience, professional supervision, development of a theory of counseling, and useful feedback are the requisites for competent counselors. Similarly, no claim can be made that this book exhaustively lists all the basics of counseling. *The Elements of Counseling* is designed to function as a brief reference tool and to provide a simple conceptual framework for thinking about counseling. We hope that what the book lacks in comprehensiveness it atones for in simplicity, clarity, and brevity.

The Elements of Counseling has many coauthors. These include past teachers, supervisors, colleagues, and students—the persons who taught us about the elements of counseling. Thank you all. Special thanks go to Drs. Thomas Frantz and Jenifer Lawrence for many of the ideas described in Chapter 4 on grief and trauma. We also acknowledge the helpful work of our UB graduate assistant, Brian Amos, and the book's reviewers:

Denise Arnold, Pierce College
Stuart Itzkowitz, Wayne State University
John Thoburn, Seattle Pacific University

Scott T. Meier
stmeier@buffalo.edu

Susan R. Davis
drsuedavis@aol.com

About the Authors

SCOTT T. MEIER is a licensed psychologist who received his Ph.D. in counseling psychology from Southern Illinois University, Carbondale. He is a professor and a faculty member in the Program in Counseling and School Psychology, Department of Counseling, School, and Educational Psychology, SUNY Buffalo.

SUSAN R. DAVIS is a licensed psychologist who received her Ph.D. in clinical psychology from Southern Illinois University, Carbondale. She is in full-time private practice.

Setting the Stage

CHAPTER I

Counseling Process

Process influences outcome. The process of counseling—what the counselor and client do in session—affects the outcome, the success of counseling (Beutler & Harwood, 2000; Orlinsky & Howard, 1978). To master process, beginning counselors must develop a repertoire of helping skills as well as a theory of counseling that directs their application.

How do you go about developing an approach to process? It is no easy task: A recent estimate placed the number of different counseling approaches (such as psychoanalytic, behavioral, and rational/emotive) at more than 400 (Kazdin, 1994; Perez, 1999). Research on psychotherapy outcome shows no broad superiority for any one approach (Lambert & Cattani-Thompson, 1996; M. Smith & Glass, 1977), and no theoretical consensus has yet been negotiated among the major schools.

Needless to say, many contemporary counselors describe themselves as integrative and eclectic. At its best, eclectic counseling involves *doing what works*. Given their training in a wide variety of counseling approaches, eclectic counselors rationally and intuitively select an approach based on the individual needs of their clients. In some cases, this judgment process is straightforward. For example, if a client appears at your office with a simple phobia, you could employ systematic desensitization with a strong expectation for success (Wolpe, 1990). The match between client and approach, however, is often unclear, and in these instances, the weakness of eclecticism is exposed. At its worst, then, eclectic counseling involves guessing or "flying by the seat of your pants."

In this chapter, we describe 13 approaches to counseling that we have found to be important, based on a mixture of research and counseling practice. Counselors use these approaches to begin and strengthen a helping relationship, and students who learn these steps will have acquired a repertoire of basic helping skills.

1 MAKE PERSONAL CONTACT

The foundation of counseling is the relationship between counselor and client. Different approaches emphasize the counseling relationship to varying degrees, but all practitioners understand that the client and the counselor must first make contact.

Because counselors lack a precise method for describing relationships between people, *making personal contact* is difficult to describe. We have to talk around the topic: Making contact means being with the client, touching someone emotionally, communicating. This does not mean the client must immediately develop an intense relationship with the counselor. In fact, the contact may need to be moderate for clients who are afraid of intimacy and personal contact.

Pay particular attention to making contact with your clients during the first session (see Adams, Piercy, & Jurich, 1991; Gunzburger, Henggeler, & Watson, 1985; Odell & Quinn, 1998; B. Pope, 1979; Sullivan, 1970). Be open to your clients' lead. If they start to chat, chat for a minute, and then return to the business of counseling.

> **COUNSELOR:** Hi, I'm Susan. I'm your counselor.
>
> **CLIENT:** Hello, I'm Bill. It certainly is snowing hard out there.
>
> **COUNSELOR:** Yes, it is. Did you have any trouble getting here?
> *(contact!)*

Not

> **COUNSELOR:** Let's talk about your problems, not the snow.
> *(too direct)*

Allowing your clients to lead in the initial stages of counseling encourages the development of trust. It also provides information about their agenda and their interpersonal styles (MacKinnon & Michels, 1971).

What could stop you from making personal contact? For example, we begin to feel rushed if we see too many clients consecutively. To avoid this, we set aside 10 to 15 minutes between clients to take notes from the previous session and review notes in preparation for the next hour. Resting between sessions enables us to attend to each client as a unique individual.

2 DEVELOP A WORKING ALLIANCE

Making personal contact is the first step in developing a working alliance, a factor often associated with continuing counseling and good outcome (Greenson, 1965; Jennings & Skovholt, 1999; Samstag, Batchelder, Muran, Safran, & Winston, 1998; Sexton & Whiston, 1994; Stiles, Agnew-Davies, Hardy, Barkham, & Shapiro, 1998; Zetzel, 1956). The task of the counselor is to engage the client in such a way that both are working together to resolve the issues that brought the client to counseling. Such alliances do not occur, for example, when the counselor attempts to force the client to change or when the client is unmotivated.

Counselors invite their clients into this working alliance by extending understanding, respect, and warmth. Thus, counselors must be skilled interpersonally.

> **CLIENT:** I'm at the end of my rope. I'm so frustrated!

> **COUNSELOR:** This is *really* a difficult time for you.

> **CLIENT:** It feels good to hear you say that. None of my friends understand what I'm going through.

Counselors are skilled listeners. By learning about clients through attentive listening and offering acceptance of clients as they are, counselors develop a bond of trust and support. Without this alliance, many clients are unable to change.

3 EXPLAIN COUNSELING TO THE CLIENT

Researchers call this *role induction* (Garfield, 1994; Hoehn-Saric et al., 1964; Mayerson, 1984; Orlinsky, Grawe, & Parks, 1994; Walitzer, Dermen, & Conners, 1999). Clients frequently approach counseling with misconceptions about the process. For example, they may expect counseling to resemble a visit to a medical doctor: diagnosis, prescription, cure. If mistaken expectations are ignored, clients may drop out or fail to make progress.

> **CLIENT:** Doc, why aren't you asking me more questions about my mental illness?

> **COUNSELOR:** John, I see counseling being most helpful when you're talking with me about your feelings. That's the best way for both of us to learn about what's going on with you.

> **CLIENT:** Well . . . okay.

In this example, the client expected the counselor to lead the counseling session by asking questions. The counselor explained, in essence, that it was the client's job to talk and that it was the current task of counseling to explore feelings. In addition, for some client problems you may also be able to provide an estimate of the likely length and potential effectiveness of counseling (Beutler & Harwood, 2000).

Beginning counselors should avoid giving explanations until they feel comfortable and knowledgeable enough to do so. Practice first in role plays. What to explain depends on factors such as the presenting problem or agency policies. Clients may find it helpful, for example, to know that (a) they will do most of the talking, (b) they may experience painful feelings before they begin to feel better, (c) exceptions exist regarding the confidentiality of counseling, (d) persons in counseling are not inherently weak, and (e) most individuals in counseling are quite sane.

Clients may also find it helpful to know that they can take some time to find a resolution to their problems. Many clients approach their first session hoping to find an immediate solution.

> **CLIENT:** I'm glad we could meet today. I'm going home this weekend and my father wants to know what my major is and what kind of job I'm going to get when I graduate.

> **COUNSELOR:** Sounds like you're under pressure to make a quick choice.

> **CLIENT:** Sort of—he keeps asking me about it and I just want to be able to tell him something.

COUNSELOR: We might be better off spending more than 50 minutes deciding on your major and your career. It's important to explore your abilities and interests as well as job requirements.

CLIENT: I guess so.

COUNSELOR: It might also make sense to talk about how your father will react to your decisions.

CLIENT: That's true.

In this example, the counselor provided the client with permission to take more time as well as a rationale for why the extra time might be useful.

J. S. Abramowitz (2002) noted that providing an explanation and rationale for counseling process and procedures may be particularly important for clients who will complete difficult treatments such as exposure therapy. He maintained that clients "who have a conceptual model for understanding their own difficulties, as well as knowledge of how and why exposure works to decrease these problems, will be much more likely to comply with the often-difficult treatment instructions" (p. 22).

Role induction may be part of a set of procedures designed to increase clients' expectations for therapeutic gain (J. Frank, 1971). Clients' (and therapists') belief in the effectiveness of counseling affects its outcome (Orlinsky et al., 1994; Rosenthal & Frank, 1958; see also Orlinsky & Howard, 1978, pp. 300–304). Part of the counselor's job, then, is to provide clients with realistic hope for improvement.

CLIENT: I'm getting so afraid of snakes that I won't leave the house much anymore.

COUNSELOR: The approach we're going to use—called systematic desensitization—has proven to be highly effective for eliminating phobias like your fear of snakes.

CLIENT: Okay . . . what do we do?

In this example, the counselor described the counseling procedure and its past effectiveness. However, once you become skilled at explaining counseling to your clients, you may be tempted to explain *everything*. This places the counselor in the foreground and encourages clients to hide behind intellectualizations.

CLIENT: After we argued I went home and kicked the dog. Is that reaction formation or displacement?

COUNSELOR: Displacement. Instead of expressing your anger to your boss you expressed it to your dog.

CLIENT: Right. I thought it was displacement, but I wasn't sure.

Here the counselor has taken on the position of expert and allowed the client to remove the focus from herself. As an alternative, the counselor might inquire how that knowledge (of classifying "kicking the dog" as reaction formation or displacement) would help the client.

Counselors also consider it an ethical obligation to employ and explain informed consent forms when beginning with new clients. Counselors often ask clients to sign consent forms that explain elements such as the terms of participation in psychotherapy (for example, a session is 50 minutes long), payment of the counselor's fee (for example, through insurance), and the client's right to terminate treatment.

Counselors may also use consent forms to describe their responsibilities (and the limitations thereof). For example, the counselor may describe her or his legal and ethical responsibilities to client confidentiality and to providing counseling of good quality. The counselor may describe the limits of confidentiality (for example, in life-threatening situations or when there is a need to consult with other counselors) and the possible effects of the counseling process for the client (for example, that the client *may* experience periods of upset as she or he examines issues).

Different professions and agencies may have different requirements regarding what should be explained to the client (for example, see Article 10.01, www.apa .org/ethics). Informed consent means that the client has:

1. Adequate mental capacity to provide consent
2. Obtained sufficient knowledge about the procedure(s)
3. Not experienced any coercion or pressure, either direct or indirect, during the decision-making process
4. Documented the decision (for example, by signing a form)

Consent can become particularly complicated if the client is receiving services that are reimbursed by a managed care company (see Davis & Meier, 2000). Such insurers may require subscribers to sign a blanket consent form when they initially purchase their health insurance. Any information provided by the client may then be shared with managed care employees as well as persons outside the company, such as the client's physicians. Counselors are well advised to remind such clients about the kind of clinical information that will be transferred to the company. An additional safeguard is to ask clients to complete the counselor's own consent forms to be certain that they fully understand and agree to any necessary procedures or sharing of information.

4 PACE AND LEAD THE CLIENT

Pacing and leading refer to how much direction the counselor exerts with the client. When pacing a client, a counselor follows along in terms of the client's expressed content and feeling.

> **CLIENT:** I put off studying until the last minute again, *(sighing)* and then I just hit the books the whole night.
>
> **COUNSELOR:** You *crammed* again.
>
> **CLIENT:** Yeah. That's what happened.

Here the counselor paced the client by succinctly restating the client's concern. Nothing was added to the client's meaning; no direction (that is, leading) was given. Pacing lets the client know that the counselor is listening and understanding.

Reflection of feeling and *restatement of content* (Egan, 2001) are two effective methods of pacing clients. Reflection of feeling refers to the counselor's recognition of the client's feeling and a subsequent mirroring of that feeling. With restatement of content, the counselor notices the client's thoughts and restates that content.

Reflection of feeling and restatement of content build rapport between counselor and client by developing a consensus about what the client is thinking and feeling. For many people, having another person listen deeply makes those experiences more real. In fact, some clients seem unaware of what they are experiencing until they hear a restatement or paraphrase.

> **CLIENT:** I was so mad I just got up and left.
>
> **COUNSELOR:** You were really angry with your partner.
>
> **CLIENT:** Yes . . . I guess I was. I just realized how angry I was—and still am!

Pacing and leading are other ways of discussing *timing*. Through experience and watching the individual reactions of clients, counselors develop a sense for when they should direct or lead the counseling process.

> **CLIENT:** I'm stuck. I just can't keep putting off studying like this.
>
> **COUNSELOR:** I wonder if your cramming has anything to do with your desire to avoid responsibility for failing the course.
>
> **CLIENT:** Well . . . my dad thinks I get Cs because I party like he did in college, not because I don't have the ability. Hmmm. . . .
>
> **COUNSELOR:** *(silence)*

In this example, the counselor intervened at a point when she believed the client was ready to hear a new message. Once the client made the connection, the counselor backed off and allowed the client to think about the new material. In general, counselors begin with much more pacing before they lead the client.

Psychologically sophisticated clients progress more quickly because of their greater familiarity with the counseling process. Highly motivated clients can better tolerate the pain and ambiguity often present during the course of counseling. Strong rapport also signals the counselor to move ahead on client issues.

> **COUNSELOR:** We've spent several sessions speaking about how you can adjust to your illness . . . but we haven't spoken about death.
>
> **CLIENT:** I know.
>
> **COUNSELOR:** *(silence)*
>
> **CLIENT:** I know . . . I know I have to start talking about dying.

The counselor recognized this individual's self-awareness of his fear of death and gently brought it to his attention. The client's awareness of the importance of talking about that fear helped him to begin the process.

Pacing occurs most naturally when client and counselor are similar on variables such as socioeconomic status, personal values, cultural background, and life experiences. Similarity encourages understanding based on intuitive insight into the other's experience. The effectiveness of peer counseling (Varenhorst, 1984), group counseling (Yalom & Leszcz, 2005), and paraprofessional counseling (Christensen & Jacobson, 1994) may be partially due to the pacing that naturally occurs in such homogeneous settings. The status of professionally trained counselors may increase credibility, but at a possible cost to their capacity to match and pace clients.

An important contraindication for leading is your state of stress: Stressed counselors may be too inclined to take risks and push their clients (Brady, Healy, Norcross, & Guy, 1995; Freudenberger, 1974; for an introduction to the area of impaired professionals, see Kilburg, Nathan, & Thoreson, 1986). In general, when counselors lead too much, they lose their clients, both figuratively and literally. Introducing three or four major insights during one session may seem like brilliant counselor strategy, but clients are likely to remember only one or two themes. Leading clients into a sensitive area too soon can be risky. One beginning counselor felt he had zeroed in on his client's major issues and began to share his interpretations during the first session. He assumed the client would recognize the truth of these insights and change. Instead, the client became frightened and dropped out of counseling.

5 SPEAK BRIEFLY

In general, counselors should talk less than their clients. Except when summarizing, communicate in one or two sentences. Unfortunately, even the best counselors can get carried away.

> **CLIENT:** My dad got angry at me, so I went to the store and got something to eat.

> **COUNSELOR:** So your father lost his temper again, and as usual, you got a stomachache when you started to feel anxious about your dad's yelling. I wonder if you were eating as a way to manage your anxiety or whether you just needed to get out of the house? Is there something else you can do to change that the next time your dad becomes angry?

In this example, the counselor shared all of her interpretations and solutions at once. If you were the client, how would you feel under this bombardment of questions?

Beginning counselors typically have difficulty being brief with talkative clients. After a goal or theme has been established (for example, to discuss the client's relationship with his father), it's okay to bring a wandering client back to the main issue.

> **CLIENT:** So when my dad yelled at me for the third time that day, I really started to feel anxious. I went down to the corner market and bought something to eat so my stomach wouldn't hurt so much. I didn't want to study, so I called my boyfriend from the store phone and we got into an argument. I haven't talked to him since then but I'm wondering if—

> **COUNSELOR:** Could we stop a moment and talk about your feelings about your father?

One technique counselors employ to show they are listening and yet stay out of the client's way is to use *minimal encouragers* (Egan, 2001). Minimal encouragers are phrases like "uh-huh" and nonverbal gestures like head nods.

> **CLIENT:** I know I have to start talking about my illness.

> **COUNSELOR:** Yes.

> **CLIENT:** And it's hard to think straight with the pain. It hurts my family to see me in pain.

> **COUNSELOR:** *(nodding)*

> **CLIENT:** They want so badly to help me that we haven't even begun to say good-bye.

This counselor's use of minimal encouragers appears effortless. Nevertheless, the client may be benefiting a great deal from exploring personal issues without counselor interruption.

6 WHEN YOU DON'T KNOW WHAT TO SAY, SAY NOTHING

Particularly in counseling, silence is golden. Perhaps because silence can indicate awkwardness in social conversations, beginning counselors often feel uncomfortable with silence and quickly fill the gaps between client statements.

> **CLIENT:** So I got laid off after 15 years of service.

> **COUNSELOR:** Fifteen years of hard work . . .

> **CLIENT:** Yeah . . . (*silence*)

> **COUNSELOR:** And you're pretty angry at the company too, I imagine.

The client might have been feeling angry but could also have been experiencing sadness about the job loss or happiness about job accomplishments. The counselor's premature statement, however, effectively stopped the client's internal processing of thoughts and feelings. Learn to notice when people are absorbed and dealing with their issues—a sign of real work in counseling.

Many clients report that during silent moments, they are processing their internal experiences and essentially taking a pause from the conversation of therapy. Levitt (2002) identified three types of pauses: (a) emotional, where clients attend to strong feelings, (b) expressive, where clients sought labels or language to describe what they were experiencing, and (c) reflective, where clients sought meaning or made connections of what they were experiencing. Levitt and Rennie (2004) indicated that reaching such silences should be a major goal of therapy "because they lead to a narrative that reflects a deepened awareness of self" (p. 303). Consequently, therapists should exercise caution during client silences so as not to prematurely interrupt this useful processing.

On the other hand, Levitt and Rennie (2004) noted two types of pauses that were not helpful to clients. In the first, termed obstructive, clients avoid processing material that would help them progress in therapy; this sometimes occurs when clients feel a topic is too dangerous. The second type is an interactional pause as clients began thinking about something confusing or troublesome their therapist had said. Levitt and Rennie provided an example of a cognitive therapist who asked a depressed client what her friends might tell her to raise her self-esteem. The client believed that her friends had in fact abandoned her when she became depressed, but thought such a response would not satisfy the therapist's question. The client became silent and eventually answered, "I don't know." Levitt and Rennie indicated that this was an example of deference by a client toward a therapist. While such deference may be meant to protect the therapeutic alliance, it also stops clients from more useful processing.

Apart from the initial interview, it is largely the client's job to talk, not the counselor's. The counselor may sit quietly during silent periods and wait for the client to resume speaking; silence here is not withdrawal. The best therapists, like good referees in sporting events, work in the background. Referees and therapists tend to be most noticed when they are making mistakes.

> **CLIENT:** After I failed that exam I was so angry that—
>
> **COUNSELOR:** You wanted to drop out of school again.
>
> **CLIENT:** —uh . . . no, I wanted to try twice as hard next time.

Not only did the counselor in this example interrupt, but he was wrong; and the former is more serious. Counselors think of themselves as helpers. Listening helps, too—sometimes it's all the help clients need.

7 YOU MAY CONFRONT AS MUCH AS YOU'VE SUPPORTED

Confrontation or challenging in counseling does not mean opposing the client but pointing out discrepancies between clients' goals and their actions (Egan, 2001). In a way, confronting the client is a way of saying, "Stop a minute! Look at what you're doing." Beginning counselors typically find such challenging difficult. A good rule of thumb is that you can confront as much as you've supported.

> **CLIENT:** I just don't have the willpower to eat more.
>
> **COUNSELOR:** I'm puzzled by that, Joe. When you started counseling two months ago, you said that you really wanted to get back to your normal weight. That sounded like "willpower" to me.
>
> **CLIENT:** Well . . . I do want to gain weight. I just can't do it.
>
> **COUNSELOR:** What's stopping you?

Support and empathy are the foundation upon which the counseling relationship is built. Consequently, confrontation is unwise during the early stages of counseling. Once you establish a bond, however, confrontation may increase client self-awareness and motivation to change (Egan, 2001).

8 IF YOU WANT TO CHANGE SOMETHING, PROCESS IT

At this chapter's beginning, we described *process* as what occurs during the counseling session. Counselors also employ the word *process* to describe the act of *talking about* something that is happening in the session. Thus, processing refers to discussion of client and counselor feelings about an event or an aspect of the counseling relationship (Cashdan, 1988).

Counselors contrast *process* with *content* (Orlinsky et al., 1994; B. Pope, 1979), the latter referring to the overt topic of the conversation.

> **CLIENT:** The problems in my marriage are all my wife's fault.
>
> **COUNSELOR:** She has . . .
>
> **CLIENT:** Started a job. She's not spending any time with me or the kids.
> *(content)*

The content in this example has to do with the client's belief that his wife is the cause of his problems. In contrast, a process comment often focuses on immediacy—that is, what the client and counselor are *feeling and experiencing at that moment.*

> **CLIENT:** I don't know what else to say.

> **COUNSELOR:** Seems that it's difficult for you to talk with me. *(process)*

> **CLIENT:** Yeah. My wife told me I should see you.

> **COUNSELOR:** So you're not here because you want to be. *(process)*

> **CLIENT:** Yeah . . . we've been fighting a lot. I'm feeling kind of shaky about what's *(sighing)* going to happen.

> **COUNSELOR:** Tell me more about that "shaky" feeling.

> **CLIENT:** I don't know. I love her, but I'm really upset about this work thing.

> **COUNSELOR:** Are you feeling shaky right now? *(process)*

> **CLIENT:** Yes . . . I don't know what else to do to make it work.

In this example, the counselor pointed out the salient aspect of the process: The client wasn't talking. By openly acknowledging the client's discomfort, the counselor helped the client to relax and open up.

If you want to focus on process, allow time for processing to occur. Processing usually cannot occur in a useful manner during a constrained time period (for example, 10 minutes before the session ends). Processing typically cannot be forced, but occurs when the counselor notices an opportunity to allow a focus on process. Once processing has occurred in counseling, counselor and client typically experience greater emotional closeness that allows for further exploration of client issues and vulnerability on the part of the client.

9 INDIVIDUALIZE YOUR COUNSELING

Each counselor eventually develops a personal style of counseling. This style may be based on a particular theoretical approach or on experience gained in a particular work setting. Remember, however, that you must adapt those general rules and personal techniques to each client (compare to Martin, 1990).

Explaining exactly *how* to individualize your counseling approach is very difficult because little consensus exists among counselors on this subject. Counselors tend to build personal frameworks for gauging how best to modify their counseling approach. Some counselors assess the psychological sophistication and level of motivation of their clients; others observe social maturity and intelligence. If the client has had previous counseling, you may be able to tailor your approach on the basis of what worked and what didn't.

Counselors observe clients' use of language in an effort to conduct sessions at a matching conceptual level. Be especially careful to avoid jargon that would confuse clients.

> **CLIENT:** I've been really down lately—I'm getting less pay because my hours at work were cut during winter. I've had to stop jogging, too, because of all the snow we've had.
>
> **COUNSELOR:** You're feeling down.
>
> **CLIENT:** Yeah . . . it's depressing.

Not

> **COUNSELOR:** I suspect your decreased rate of receiving positive reinforcement is directly related to your depressive feelings.

Individualize you must. Often it is the counselor who must adapt because many clients begin counseling exhibiting inflexible behaviors. As the counselor, you may alter your language, your posture, even which counseling approach you employ. By doing so, you will create a unique application of your counseling style for each client. Remember, you must talk *with* your clients, not *to* them.

10 NOTICE RESISTANCE

Resistance refers to an obstacle—presented by the client—that influences the process of counseling (Wachtel, 1999). Resistance may be a key indicator of the client's readiness for change and the types of interventions the counselor should employ.

Counseling theorists construe resistance differently, and most acknowledge the inevitable presence of resistance. Psychodynamic counselors, for example, view resistance as an attempt to keep anxiety-provoking material from awareness. For some clients with a traumatic incident or difficult situation, discussion about the distressing topic can be analogous to opening a door to a room full of flames. Social learning counselors see resistance as stemming from fear of the consequences of changed behavior (Hansen, Rossberg, & Cramer, 1994). Mahoney (1987) suggests resistance to change "is a natural expression of self-protection" (p. 15) when clients' ways of thinking, feeling, and behaving are challenged.

All of these ideas relate to clients' experience of psychological safety. In this context, resistance is something to be respected, understood, and, when appropriate, explored.

> **COUNSELOR:** Gloria, last week we agreed you would speak with your mother about some of the issues we discussed. But you haven't mentioned anything so far today about that.
>
> **CLIENT:** Oh . . . did I say I'd speak with her? I guess I forgot.

For some clients, it simply takes some time before they are ready to talk about difficult experiences with their parents, family, or friends. Recent research suggests that adolescent clients may underreport, at the beginning of counseling, such socially undesirable behaviors as smoking, cheating, and failing to follow directions from adults (Meier & Schwartz, 2007, cited in Meier, 2008). This research found that adolescents increased their reports of undesirable behaviors as counseling progressed, suggesting that these clients provided more honest information as they developed a working alliance with their counselors over time.

Other examples of resistance include an abrupt change of topic and forgetting important material. If the client continues to make progress despite such behavior, simply make a mental note of the resistance and go on. However, if you encounter repeated resistance, process it with the client at a time that seems right to you and at an emotional intensity that fits the client. This discussion may be all that's necessary for the client to move forward.

> **CLIENT:** Oh . . . did I say I'd speak with her? I guess I forgot.

> **COUNSELOR:** That's strange—I know you really wanted to talk with her at the end of last week's session.

> **CLIENT:** I guess I did.

> **COUNSELOR:** How do you feel as we talk about it now?

> **CLIENT:** Well . . . scared, really. I realized how angry she'll be if I confront her about abusing me while I was a kid.

> **COUNSELOR:** You're frightened about confronting her.

> **CLIENT:** Yes . . . but I want to do it. That seems like the only way for me to resolve it.

With *extreme resistance*, the counselor returns to pacing the client by decreasing the emotional intensity of the session or by changing the topic (Brammer & Shostrom, 1989). In essence, the counselor gives the client permission to deal with the intense material when the client is more able to do so.

11 WHEN IN DOUBT, FOCUS ON FEELINGS

No matter what their theoretical orientation, counselors often focus on the feelings of clients (Merten, Anstadt, Ullrich, Krause, & Buchheim, 1996; B. Pope, 1979; Walborn, 1996). Counselors trust clients' feelings—particularly as expressed on the nonverbal level—as indicators of salient issues (Angus & McLeod, 2004; Greenberg & Angus, 2004).

> **COUNSELOR:** I have a sense that it's really important for people to like you.

> **CLIENT:** Yes—isn't it for everyone?
> *(tensing)*

This client responded defensively, interpreting the counselor's statement to mean that such a need was abnormal or inappropriate. The counselor might follow up by asking the client what he feels when he wants people to like him.

Learning how to recognize and express feelings challenges many clients (Cormican, 1978). At one extreme are clients who treat their feelings around difficult or traumatic situations as if those feelings were a room full of flames ready to leap out the second someone opens the door. Other clients may begin counseling unable to recognize their feelings or to describe them in more than a cursory manner (Barrett, 2006).

> **COUNSELOR:** How do you feel about losing your job?

> **CLIENT:** Bad. I'm very upset.

Although this client can label the intensity and direction of the feeling, descriptions such as "bad/good" provide very little detail. When clients use nondescript words such as *bad, good,* or *upset,* ask them to elaborate. The client in the preceding example might be feeling angry because she has to go through the job-hunting process again or sad because she believes that the job loss indicates failure on her part.

Counselors typically search for the *Big Four* of feeling words—(1) *anger,* (2) *sadness,* (3) *fear,* and (4) *joy*—which some research suggests are biologically based and evolutionary in origin (Izard, 2007). Beginning counselors are often taught to help clients recognize these feelings and the reasons for them; for example, I *feel angry because* I didn't get the job (Egan, 2001). An ability to recognize these feelings in clients (and to help clients become aware of them) is a sign of progress in the beginning counselor.

Counselors also attend to feelings because clients seek counseling primarily to alleviate psychological pain. Helping clients pay attention to their feelings can increase their motivation to change. Some research suggests that a moderate level of emotional arousal is most helpful for motivating clients to change in therapy; thus, counselors may try to increase the affective focus for clients who present with little emotional stress (Beutler & Harwood, 2000). In addition, although research indicates that emotional suppression can produce negative cognitive and physiological effects (Moses & Barlow, 2006), it is quite common for clients to suppress what they experience as an overwhelming feeling. Fully experiencing feelings may bring insight and relief. Some counselors work to create a *corrective emotional experience* in which the major goal is to provide the client with an affirming, safe, and close therapeutic relationship that undoes maladaptive interpersonal learning that occurred in a previous relationship (Teyber, 2005).

Experienced counselors can determine clients' feelings by paying attention to how *they* feel as clients talk. By sharing their reactions to clients' situations, counselors can increase clients' experience of their feelings in the here and now.

> **CLIENT:** Even though it's been a year since my father died, I still feel a hole in my life.
>
> **COUNSELOR:** As you spoke about your father, I felt sad.

This technique may help a client who is having difficulty experiencing or recognizing feelings. The counselor models the feeling and reassures the client that such a feeling is okay.

A few clients see expression of feelings as an all-or-nothing proposition. These clients follow the volcano theory of emotion: They express no affect until their feelings build up and explode.

> **CLIENT:** Jim, my coworker, just kept teasing and teasing me until I couldn't stand it anymore and I screamed at him.
>
> **COUNSELOR:** You were feeling very angry. How did he react when you yelled at him?
>
> **CLIENT:** He was surprised.

Assertiveness training, where clients learn how to express feelings to others, may be appropriate for such clients (Ballou, 1995; Eisler, 1976). Clients may also learn to express emotions at a moderated level by practicing such expression in counseling.

12 PLAN FOR TERMINATION AT THE BEGINNING OF COUNSELING

Termination refers to the process that occurs at the end of counseling (Anthony & Pagano, 1998; Kane & Tryon, 1988; Perlstein, 1998; Sexton, Whiston, Bleuer, & Walz, 1997). At the beginning of counseling, client and counselor should reach at least a tentative understanding about when and how counseling will finish.

COUNSELOR: When will we know counseling is complete?

WIFE: When we're not fighting.

HUSBAND: At least not as often as we are now, which is daily.

COUNSELOR: If you fight only once a week, then counseling has worked?

BOTH: Yes.

Planning for termination means that explicit goals have been set, although this does not necessarily mean that client and counselor have created a laundry list of goals (see Cormier & Cormier, 1998, for specific information about goal setting). Goals should become clearer and may be revised (along with the termination plan) as clients move deeper into self-exploration (Meier, 2003). A plan for termination can be part of treatment planning (see Chapter 4), the steps counselor and client will take toward successful resolution of the client's problem.

CLIENT: I don't want to be just emotionally independent of my family—I want to have my own apartment too.

COUNSELOR: Moving out would signify a real step toward independence for you.

CLIENT: Yeah. If I did that, I'd be satisfied.

Termination should be planned, not abrupt; your comfort with termination will influence this process. For some beginning counselors, termination can be frightening when they realize that counseling will be concluded and clients sent on their way. As an aware graduate student put it (J. Englert, personal communication, April 4, 2004):

> With termination rapidly approaching, I find that each time I remind my clients of how many sessions we have left, I am in a sense reminding myself that I have to face all of my clinical fears head on. As scary as it was starting with clients, it is even more frightening to finish the relationship and send them on their way. I feel as though I will have to face any inadequacies I had through our work during termination, and will have to feel strong enough to keep the termination sessions in control.

Thus, beginning counselors may fear that their mistakes and inadequacies will be most apparent during the termination process. If you have similar feelings, supervision is the best place to address them.

Overall, remember that it is important to say good-bye, consolidate the counseling experience, discover what counseling meant to the client, and discuss future situations.

CLIENT: So it looks like I'm going to go to college after all.

COUNSELOR: I'm delighted. You worked very hard to sort things out with yourself and your family.

CLIENT: I certainly appreciate your help and listening to me while I was so confused about what to do.

COUNSELOR: Well, I enjoyed listening to you. I'm glad you're ready to go, but sad we won't be meeting anymore.

CLIENT: Me too . . . but I know I can talk things out with my parents in the future.

Client issues may also appear or reappear during the termination process. Past grief may surface, or the client may bring up a major event that was never previously mentioned (for example, sexual abuse). But termination also presents special opportunities: The client or counselor may use this time to sum up progress and achievements, to describe what could be worked on in the future, and to discuss possible referrals. Regarding the last point, in settings such as schools and college counseling centers, the end of an academic year or graduation may mean the end of the counselor's work with a particular client, but not the end of the need for counseling for that client.

Clients say good-bye in different ways. Some persons cling to the relationship and try to avoid ending; some deny any feelings; some cancel the last session. With clients who terminate prematurely—even though they may be satisfied with counseling—no formal termination may be possible. Talk with clients about how they feel about terminating. All this can be made simpler by establishing at the onset of counseling the conditions under which counseling will end. Counselors often schedule several sessions to terminate. With some clients, they may schedule a monthly meeting to maintain progress.

13 ARRANGE THE PHYSICAL SETTING APPROPRIATELY

The counseling process is affected by the physical characteristics of the counselor and the counseling setting. (See Langs, 1973, and Sommers-Flanagan & Sommers-Flanagan, 2002, for more complete treatments.)

A. DRESS APPROPRIATELY

Dressing for success may be more necessary in business than in counseling, but some clients judge counselors by their attire. An appearance that matches that of other colleagues is fitting. Dress becomes problematic when counselors draw attention to themselves through apparel.

B. ATTEND TO PHYSICAL SPACE

Observe how clients use space. Some clients may prefer that there be no physical objects between you, whereas others may feel safer with a barrier (such as a desk or small table). We prefer to sit facing clients, although some counselors have suggested that sitting at an angle increases client comfort. If possible, provide comfortable chairs for counselor and client. This makes it easier for both to concentrate during the entire length of the session.

c. CONDUCT COUNSELING IN A QUIET SETTING

If you can overhear another conversation during a counseling session, it's likely that someone can overhear your (confidential?) discussion. Respecting clients' privacy and confidentiality means counseling should occur in a quiet, private setting.

d. AVOID INTERRUPTIONS AND DISTRACTIONS

Don't accept phone calls; don't read the mail; don't finish your lunch. Attention in a counseling session belongs exclusively to the client, except in the case of an emergency.

e. BE PROMPT

If clients arrive late or leave early, that's an issue to address in counseling. In any event, clients should not be wondering about the whereabouts of their counselors 10 minutes after the scheduled start. A regularity to the beginning and end of counseling sessions signals that clients can trust this relationship. Similarly, arranging the session for the same time each week or day can signify the counselor's commitment to the relationship. Some clients measure their importance to the counselor by how often their sessions exceed the scheduled time. Be explicit with clients about the length of a session. A typical individual counseling session lasts 45 or 50 minutes. This is enough time to go beyond the superficial but not so long that client and counselor are exhausted.

f. INVEST IN A BOX OF TISSUES

Sometimes clients cry. It's generally not a good idea, however, for the counselor to hand a tissue to a client. It may imply, to some clients, that it is time to stop being sad. Let clients use tissues when they're ready.

g. REMEMBER CONFIDENTIALITY

Restrict your communications about clients to professional colleagues in professional settings. Counselors may be tempted to discuss clients in public places, such as restaurants, teachers' rooms, or informal gatherings. Particularly in small towns and other settings where people who overhear those conversations might know the individuals involved, talking about clients in public has the potential for serious trouble. Restrict your discussion of clients to supervisors and necessary colleagues in work settings. Also, find out whether your state or specialty has specific legal and ethical guidelines about confidentiality. Most professional organizations discuss confidentiality and other ethical issues on their websites (for example, www.apa.org, www.aamft.org, www.counseling.org, www.naswdc.org, www.nbcc.org, and www.psych.org).

SUMMARY AND DISCUSSION QUESTIONS

Conversations between counselor and client are different from those ordinary exchanges we have with our family and friends. Counselor-client conversations are therapeutic in intent, aimed at helping the client change, and progressive in nature. Process comments by the counselor also distinguish counseling from normal conversations.

This chapter described the first phase of counseling in terms of what the counselor can do to help the client begin the counseling process. The relationship begins and a bond develops, often signaled by an emotional closeness. The client begins to get a sense of what counseling is about and the counselor a sense of who this person is and how quickly she or he can progress.

Another important way the counselor-client conversation differs from ordinary discourse is the discipline the counselor brings to the interaction. In general, counselors stick to their basic listening skills until those skills become fluid and automatic. Once mastered, the counselor makes an exception to the basic form only when she or he thinks through the situation carefully. For example, a counselor typically will not share her or his personal reactions to a client without thinking about the likely effects of such a disclosure on the particular client. We know of a number of counselors who have learned this lesson the hard way. One, working as a career counselor at a high school, was interviewing new freshmen about their career aspirations. As these young students came into his office individually, the counselor heard the usual variety of careers until a girl replied, "I want to be president." The counselor was a bit confused and asked "President of what? A company?" No, the girl replied, "President of the United States." The counselor laughed out loud, but noticed that the girl was hurt. Three years later, the counselor learned that the impact of this event was still present. At this school all students complete a survey during their senior year to evaluate their experiences, including any counseling. The counselor in question was reading through these surveys when he found one that said, "Thanks for laughing at me Mr. X."

All of the skills described in this chapter are related to the counselor's ability to listen well. The counselor strives to listen well (often speaking briefly or not at all) and to help the client attend to what she or he is experiencing and feeling at this moment. It is no coincidence that many good training programs begin with a course in listening skills. As discussed in this chapter and the next, listening well makes a conversation therapeutic and sets the stage for subsequent therapeutic interventions. Listening builds relationships, even between individuals of very different backgrounds (Bruner, 2004).

To help you understand and apply the information in this chapter, consider these questions:

How do content and process differ?

What influences outcome in counseling, independent of the counseling process?

What might go wrong in the process of developing a working alliance?

What personal characteristics might enhance a client's ability to make progress in counseling?

What counselor behaviors can enhance client progress?

What counselor behaviors can inhibit client progress?

Provide an example of how you have or could individualize counseling with a particular client.

STRATEGIES TO ASSIST CLIENTS IN SELF-EXPLORATION

This chapter provides additional guidelines about a particularly important aspect of the counseling process: client self-exploration. Self-exploration refers to the elaboration and deepening of self-awareness and self-concept that occur as clients speak about themselves. Self-exploration provides information about what must be done for change to occur, and it can be therapeutic in and of itself. Once clients understand which behaviors will produce the desired outcomes, some clients change relatively quickly (see Butcher & Koss, 1978; Crits-Christoph, 1992).

Facilitating client self-exploration depends on what you *don't do* as well as what you do. Beginning counselors often work too hard: They feel responsible for *helping* clients, and they take that task as a mandate to give advice, ask scores of questions, and solve problems. In contrast, experienced counselors initially stay in the background and help clients thoroughly understand themselves and their problems.

14 AVOID ADVICE

Friends and family members give advice; counselors generally don't, particularly in the initial stages of relationship building (compare to Sexton et al., 1997). Many counselors (beginning and advanced), however, appear to believe that they must offer counsel.

COUNSELOR: What brings you here today?

CLIENT: Basically, I'm having trouble with my boyfriend.

COUNSELOR: Have you tried talking with him?

Even though the counselor's response is a question, it implies advice (go talk with your boyfriend). Advice is an intervention and as such should be avoided until counselor and client establish a trusting relationship. Advice can also be interpreted as indicating that the counselor has stopped listening to the client and is now ready to fix the client. Many counselors completely avoid advice because (a) clients have already tried these simple strategies, and (b) the advice has already failed or the client wouldn't be in counseling.

> **CLIENT:** I'm having trouble with my boyfriend.

> **COUNSELOR:** Tell me about the trouble you are having.

It's a much better strategy to encourage self-exploration, as in the preceding example, than to engage in premature problem solving. Because friends and family frequently provide advice to clients, your advice is likely to be old news. Also, some clients resist advice because it is generated externally; these individuals expect and want to solve their own problems.

Don't confuse giving information with giving advice (Cormier & Cormier, 1998; Kirschner, Hoffman, & Hill, 1994). When you give advice, you typically provide clients with specific actions to perform. Information, on the other hand, consists of knowledge, alternatives, or facts that clients may find useful in their decision making. As explained in the following section, the usefulness of providing information also depends on when it is done.

15 AVOID PREMATURE PROBLEM SOLVING

Problem solving too early in counseling usually fails (see Egan, 2001). By *problem solving* we mean efforts by the counselor to create alternatives and suggest strategies to resolve clients' problems. Clients may resist promising solutions or already know what to do; even if clients implement a counselor's suggestion, they have only the counselor to blame or praise. Clients bear the ultimate responsibility for change. Similarly, the problems clients are facing are unlikely to be their last, and a solution selected by the counselor will not help clients learn how to handle future difficulties.

> **CLIENT:** I just can't find the willpower to start studying a week before an exam.

> **COUNSELOR:** Maybe if you drew up a schedule with regular study times you could study more consistently.

> **CLIENT:** That would work, but I doubt that I could stick to the schedule.

Often, clients know how they could obtain desired goals and outcomes but have no confidence in their personal capacity to act in the required manner. These clients lack what Bandura (1977, 1997) calls *self-efficacy*, a belief that one can perform a behavior that leads to a desired outcome. For example, alcoholic clients who have low self-efficacy for stopping drinking may relapse when they are repeatedly offered drinks at a party. Self-efficacy expectations influence whether clients initiate certain behaviors and persist in the face of obstacles.

Rather than problem solve early in counseling, counselors do better to help clients define problems fully. For example, you might explore what solutions have already been tried. This may help counselor and client better understand why previous efforts have failed or have exacerbated the original problem.

COUNSELOR: You've said that you plan to study a week before the exam, but you don't follow through. What stops you?

CLIENT: I start to study—I stay up all night and read, but after a couple of nights of that I just don't feel like working anymore.

COUNSELOR: Sounds like you just get worn out.

CLIENT: That's right.

A client's solution to a problem can sometimes produce additional problems or exacerbate the original difficulty (Watzlawick, Weakland, & Fisch, 1974). In the preceding example, the client became physically fatigued because of poor study habits and lost motivation to study further.

16 AVOID RELYING ON QUESTIONS

Faced with their first real or role-play clients, new counselors may simply ask questions. Beginners use such questions to elicit more information or to suggest advice. It is certainly permissible to ask about specific information or to seek elaboration, but clients may perceive a sequence of questions as threatening.

CLIENT: So I just left after we started to argue.

COUNSELOR: You were angry?

CLIENT: Yes.

COUNSELOR: Why didn't you stay and work things out?

CLIENT: I was too angry to talk.

COUNSELOR: You felt out of control?

Questions, particularly "why" questions, put clients on the defensive and ask them to explain their behavior. Questions keep the counselor in sole control of counseling and may inappropriately lead clients. Instead of asking a question, a counselor might say "Tell me more about that . . ." Another option is to use statements that *reflect* the content or feeling clients have expressed; these statements do not imply direction by the counselor.

CLIENT: I just left after we started to argue.

COUNSELOR: You got up and left the room.

Counselors also differentiate between open and closed questions (see Adams, 1997; Egan, 2001). Open questions seek elaboration by clients; closed questions ask for specific information and may be answered in a word or two. Counselors particularly should avoid sequences of closed questions when they are facilitating client self-exploration.

CLIENT: I just left after he started to swear at me.

COUNSELOR: How did you feel then?
    (open)

CLIENT: I was really angry. He had no right to speak to me that way—we had been close friends for years.

COUNSELOR: Did you swear back?
   (closed)

   CLIENT: No.

The open-ended question in this example encouraged the client to describe her feelings and the reason she was angry. The closed question provided no pertinent information. Unless you need very specific information, stay with open-ended inquiries.

17 LISTEN CLOSELY TO WHAT CLIENTS SAY

Listening well to clients has multiple benefits. First and foremost, listening enhances empathy in the counselor. In this context, listening well promotes the development of a therapeutic intimacy, an emotional closeness that allows the development of a corrective emotional experience (Teyber, 2005) that has the potential to generalize beyond the therapeutic relationship.

As a counselor, you will primarily listen for the meaning of the words your clients use. You can, however, also listen for individual words and grammar. Words and the manner in which they're employed can provide important clues about how clients view the world (Bandler & Grinder, 1975; de Luynes, 1995; Wachtel, 1993).

   CLIENT: My children always disobey me.

COUNSELOR: Always?

   CLIENT: Well . . . not always. They usually get up and get dressed for school without any trouble.

A word like *always* indicates a distortion of reality, a predisposition to perceive different events in the same way. In the preceding example, the client didn't really mean *always* when he described his children's misbehavior. Helping the client to recognize this distortion of reality may help him perceive his children differently or discover a new solution.

   CLIENT: My children always disobey me.

COUNSELOR: Always?

   CLIENT: Well . . . not always. They usually get up and get dressed for school without any trouble?

COUNSELOR: What is it about that situation that enables them to do what you've asked?

   CLIENT: Well . . . I wake them up early enough so even if they get distracted for a few minutes, they still have enough time to get ready for the bus.

Similarly, counselors should determine what clients mean when they use words like *must* and *should*.

   CLIENT: I must get an A on this test!

COUNSELOR: There's a feeling of desperation in your voice.

   CLIENT: Yes. Without this grade, I'm doomed.

Albert Ellis, the developer of rational/emotive behavior therapy (REBT; see Chapter 6), is particularly vigilant in listening for words like *must* and *should* because he views them as signals of irrational beliefs (Ellis, 1998; Ellis & Grieger, 1977; Ellis & Harper, 1976). Ellis works to detect thinking that has no basis in reality and that tends to catastrophize events.

When clients speak, they may fail to describe fully their experience (Bandler & Grinder, 1975). What is deleted or missing may be indicative of the client's problem.

COUNSELOR: Without an A, you're doomed to what?

CLIENT: Well . . . I don't know. I won't maintain my 3.5 grade point average.

COUNSELOR: So . . .

CLIENT: I wouldn't get into med school.

COUNSELOR: And then?

CLIENT: Well . . . I'd be very disappointed. I guess I'd have to find another career.

Asking clients to present a fuller description of themselves is a method of facilitating self-exploration and helping clients change. Narrative therapists use the term *deconstruction* to describe the process of helping the client to become aware of her or his dominant story and question its elements so that it becomes just one possible view of the self (Polkinghorne, 2004). Other life events become more fully incorporated into the story and the client is placed as a more active protagonist. For example, the student who reports that she "must get an A on a particular test" may also tell stories of other tests and courses in which she feels pressure to perform extremely well. But she is also likely to be able to recall situations where she did not perform extremely well and no catastrophic consequences occurred.

18 PAY ATTENTION TO NONVERBALS

People communicate with each other by paying attention to the verbal content of messages and to the most overt nonverbal messages (for example, smiling, frowning, making a fist). But the more subtle nonverbal components of communication— tone of voice, facial expressions, eye contact, and body motion—convey equally rich information (Wiener, Budney, Wood, & Russell, 1989). The nature of nonverbal communication can be influenced by culture (Juang & Tucker, 1991; D. W. Sue & Sue, 2002), but unfortunately, there is no Rosetta stone that can be used to decipher the meaning of nonverbals with any particular client. No one can tell you exactly what to look for, but the importance of observing others' nonverbals, in role plays and everyday situations, cannot be overemphasized.

One particularly effective way that counselors confront clients is to point out discrepancies between nonverbal and verbal communication.

CLIENT: I was really upset when she said she didn't like me.
(smiling)

COUNSELOR: So you were upset . . . but I notice you were smiling when you said you were upset.

CLIENT: I was? Well . . . it is kind of hard for me to admit she upset me.

When faced with conflicting messages on nonverbal and verbal levels, counselors tend to trust nonverbal communication as more indicative of basic feelings. The assumption is that it's easier to censor verbal than nonverbal communication. Counselors often help clients become aware of the nonverbal level.

> **CLIENT:** I was really upset when she said she didn't like me.
> *(smiling)*

> **COUNSELOR:** You were upset . . . but I notice you were smiling when you said you were upset.

> **CLIENT:** I was? Well . . . it is kind of hard for me to admit she upset me.

> **COUNSELOR:** How do you feel when you smile like that?

> **CLIENT:** I don't know . . .

> **COUNSELOR:** Smile again, just like you were.

> **CLIENT:** Okay . . .
> *(smiling)*

> **COUNSELOR:** How do you feel now?

> **CLIENT:** Well, maybe a little safer . . . maybe you'll still like me even though she upset me.

This client was able to learn from exploration of her nonverbal movements. She learned that she smiled to hide her embarrassment about her feelings.

Counselors also express different messages on verbal and nonverbal levels. For example, beginning counselors sometimes experience difficulty being congruent in their communication (Shertzer & Stone, 1980).

> **COUNSELOR:** Your dad died two years ago today.
> *(smiling)*

> **CLIENT:** Yes . . . but I still feel sad about it.
> *(confused)*

Some counselors simply feel more comfortable if they smile while around others. Such habits, however, can send mixed signals to clients. Videotaping and audiotaping sessions can help you discover any idiosyncratic nonverbal expressions that interfere with the counseling process.

19 FOCUS ON THE CLIENT

Clients often talk about other people. Others may be seen as sources of one's troubles or as standards for behavior.

> **CLIENT:** My mom thinks I should go into social work.

> **COUNSELOR:** How do you feel about that?

With this client, the counselor has the choice of following the other person's feelings ("What makes your mother think you should be a social worker?") or of following the client's feelings ("How do you feel about your mother's belief?"). Almost invariably, counselors return the focus to the client (Sexton et al., 1997). The client is the object of change in counseling, not other people. On the other hand, when others may influence the process of change, as in family or marriage counseling, counselors arrange for those relevant individuals to be present.

Many clients experience difficulty in talking about themselves, and concerns about confidentiality with managed care plans may exacerbate this issue (Davis & Meier, 2000; Kremer & Gesten, 1998). Counselors can help these clients through appropriate *self-disclosures* (Jourard, 1971; Robitschek & McCarthy, 1991) or *self-involving* responses (McCarthy, 1982; McCarthy & Betz, 1978; Watkins & Schneider, 1989). With self-disclosure, the counselor reveals a personal feeling or experience of his or her own; with self-involving responses, the counselor responds personally to a statement by the client. In both cases, the intent is to help clients get in touch with *their* feelings.

> **CLIENT:** So I got laid off after 15 years on the job. Now I've got to begin job hunting, and I haven't the slightest idea where to start.

> **COUNSELOR:** I know what you mean. I lost my last job when the agency had its *(self-disclosing)* funding cut. I was depressed.

Or

> **COUNSELOR:** I feel sad that you lost your job.
> *(self-involving)*

> **CLIENT:** Well . . . I have been really down. I've been trying to think positively and ignore those feelings because I've got to get a job as soon as I can.

Self-disclosing and self-involving statements encourage clients to reciprocate. Self-involving responses in particular may help clients become aware of their feelings and speak more about themselves. Counselors' self-disclosing and self-involving statements, however, should be employed sparingly (to maintain the focus on clients) and at a matching level of intensity. As an example of the latter, it would be inappropriate for a counselor to disclose her or his current feelings of depression to a person seeking career counseling. A more appropriate self-disclosure might involve the counselor's revelation of past feelings of confusion and anxiety surrounding career decisions.

20 BE CONCRETE

Some of counseling's working materials, such as feelings and thoughts, are intangible. For example, counselors have compared feelings to the wind, suggesting that clients should notice emotions and allow them to pass through and change. Skilled counselors help clients make their feelings and thoughts concrete.

> **CLIENT:** Even though it's been a year since Dad died, I still feel a hole in my life.

> **COUNSELOR:** Tell me about that hole.

> **CLIENT:** It's kind of an emptiness . . . a numbness.

> **COUNSELOR:** You're holding your stomach as you talk about the emptiness. Hold your stomach a minute . . . tell me what you feel.

> **CLIENT:** I feel . . . sad.

Grounding feelings within the body is one method of making feelings concrete. As suggested in the preceding example, the client might learn to pay attention to his stomach whenever he feels sad. Gestalt counselors use clients' body movements, such as repeatedly making a fist or striking a pillow, to help strengthen clients' experience of feelings.

Concreteness becomes especially important when discussing clients' behavior and goals (Araoz & Carrese, 1996). What exactly do clients want to change? How will client and counselor know when counseling is finished?

CLIENT: I want to do better in school.

COUNSELOR: What do you mean by "better"?

CLIENT: I don't know . . . maybe get a 3.0 average.

COUNSELOR: You want a 3.0. Are you speaking about just this quarter or the whole year?

CLIENT: Just this quarter.

Clients sometimes have difficulty going beyond broad ideas and intellectualizations. Counselors help these clients by asking about specific events.

CLIENT: You know, if my husband would just complete the separation process with his family, we wouldn't argue so much.

COUNSELOR: Give me an example.

CLIENT: Well, last week we started to argue about the phone bill, and immediately he calls his mother for advice. Then his parents are angry at me, too, and I really feel anxious.

In this example, the counselor learned that the family of the client's spouse becomes inappropriately involved in their marriage. From such information, the counselor might decide to pursue marriage counseling or help the client plan for such events in the future.

21 UTILIZE METAPHORS

A *metaphor* is a figure of speech that contains an implied comparison—expressing an idea in terms of something else (Close, 1998; Myers, 1998; Reich, 1998; J. S. Young & Borders, 1999). Clients will occasionally offer metaphors about their issues. Listen for and accept these gifts—they are worth their weight in gold for what they reveal about how clients construe their world (compare to Martin, 1994).

CLIENT: I'm making progress in counseling, but it sure is hard.

COUNSELOR: You're making a lot of effort.

CLIENT: Yeah . . . it reminds me of a 15-round boxing match, and we're only in round 3.

Metaphors are easy to remember and help clients to think about their concerns and experiences (Cummings, Hallberg, Slemon, & Martin, 1992). By using the boxing metaphor, the counselor could ascertain the client's perception of progress in counseling. (For example, round 7 would indicate further progress than round 3.)

Another client compared her problems to a box: The size of the box indicated the client's perception of the current intensity of her problem. When the client felt she had to place her attention elsewhere (for example, to study for a test), she "put the box in the closet for now."

22 SUMMARIZE

Summarizing refers to a brief review of the major issues in counseling. You might summarize important issues during the counseling session, at the end of a session, or at the final counseling session. Client or counselor may summarize. Client summaries present counselors with an opportunity to check their ideas about the counseling process with clients.

> **COUNSELOR:** So if you were to sum up what has gone on in counseling so far, how would you describe it?

> **CLIENT:** Well, I certainly understand how my parents and I get started on arguments. Everybody gets so involved in presenting their case that no one listens to the other. I know I have to point that out to them, and notice it myself, or I'll get really frustrated in a hurry.

> **COUNSELOR:** That certainly fits with what I've heard you say in here.

On the other hand, beginning counselors may be surprised at how often a counselor's judgment that a session was useful (or not) may be contradicted by the client's perception of that same session (Borders, Bloss, Cashwell, & Rainey, 1994; Yalom & Elkin, 1990). In the following example, the counselor helps the client gain insight into her current feelings of anxiety by summarizing a theme (the client's mother protected her from taking risks throughout childhood) that had appeared several times previously.

> **CLIENT:** Any time I start a new class I feel really anxious. I just don't have any confidence in myself about taking risks.

> **COUNSELOR:** I wonder if your lack of confidence has anything to do with your mother's overprotectiveness. You've talked about how she rarely let you try new experiences on your own as a child. I wonder if the two are related . . .

> **CLIENT:** That makes sense.

Summarizing is another method of structuring counseling. Summaries often describe important themes, keep track of change in counseling, and help connect related issues.

SUMMARY AND DISCUSSION QUESTIONS

The second phase of counseling involves helping clients to deepen their understanding of relevant problems. Listening remains key, and for the beginning counselor this may mean a change from ordinary conversational patterns such as offering advice, problem solving, or use of multiple questions. Listening also involves helping clients to identify and express feelings. Counselors can help clients learn this skill through a variety of methods, including grounding feelings in the body and focusing on nonverbal expressions of affect. Deeper understanding can also be facilitated through summaries and use of metaphors, and helps prepare clients to move toward change in their presenting problem.

To help you understand and apply the information in this chapter, consider these questions:

What counselor behaviors described in this chapter inhibit client progress?

How does listening well enhance counseling process?

What is the difference between open-ended and closed questions?

Provide one or more examples of how a client communicates nonverbally.

How might the methods of qualitative research help us understand a client's use of language?

A FEW MISTAKEN ASSUMPTIONS

Counseling remains a poorly understood profession among laypersons. Many people still see counseling, for example, as a place for only the sick, weak, or severely disturbed. Beginning counselors must unlearn some of the common assumptions they bring with them to training. What counselors believe about the processes of change will influence how they conceptualize and do counseling. In a sense, this book is a role induction for beginning counselors and a role reminder for experienced therapists.

23 AGREEMENT DOES NOT EQUAL EMPATHY

Some beginning counselors interpret *empathy* to mean agreement or sympathy. Empathy refers to a deep comprehension of the subjective world of clients (see Egan, 2001). Agreement suggests that the counselor approves of the client's behavior, and sympathy indicates that the counselor feels sorry for the client.

> **CLIENT:** I was really angry when he said to go home.
>
> **COUNSELOR:** You felt angry with him.
> *(empathy)*
>
> **COUNSELOR:** Good for you. I'm glad you left!
> *(agreement)*
>
> **COUNSELOR:** He is awfully mean to you!
> *(sympathy)*

Friends and family provide agreement and sympathy. Counselors provide empathy to help clients explore their problems and become aware of their feelings and thoughts. In that way, clients begin to understand what they need to do to change.

24 DO NOT ASSUME THAT CHANGE IS SIMPLE

Clients typically expect to change some aspect of their lives so they will feel better, but matters are seldom so simple. Human behavior has multiple causes, and no counselor can always be aware of all the factors helping and hindering change. Factors such as significant others, biological influences, culture, and individual differences in response to therapy can all contribute to the complexity of the counseling process (Krause, Howard, & Lutz, 1998).

> **CLIENT:** I started on the homework, but Mother says she doesn't think it will work. She says I should stop counseling.
>
> **COUNSELOR:** Well . . . that's the first I've heard about your mother since our first *(surprised)* session.
>
> **CLIENT:** I think she might be right.

The counselor had been proceeding on the assumption that changes in the client's behavior would resolve the presenting problem. However, the client's mother has intervened and may now need to be included in the counseling process.

Counselors sometimes do marriage and family therapy because they view individuals' problems as relating to the social systems to which they belong (deShazer, 1982; Minuchin & Fishman, 2004). Change is not simply a matter of altering the "misbehaving" person. An adolescent may act out at school to stabilize her family system: Her feuding parents, considering divorce, might table their disagreements to handle the current crisis. More effective counseling would focus on the entire family rather than just on the client.

Another perspective held by counselors about the complexity of change concerns the client's readiness to change. That is, some clients may simply not yet be ready to change. In Prochaska's (1995; see Chapter 6) stages of change model, for example, clients may not be ready to change because (a) they lack awareness of the problem, (b) they are aware of the problem but are only beginning to consider how to remedy it, or (c) they are only in the beginning stages of taking actions to change.

25 POSITIVE THINKING DOES NOT EQUAL RATIONAL THINKING

Some beginning counselors mistakenly equate Ellis's concepts of rational and irrational thinking with positive and negative thinking. Positive and negative thinking usually refers to beliefs that one will encounter good or bad fortune. Irrational thinking refers to a belief, unsupported by objective evidence, that leads to a painful feeling (Ellis, 1998).

> **CLIENT:** I must get an A on all of my final exams or I'm doomed.
>
> **COUNSELOR:** You're saying you must be perfect in all of your classes.

As we noted in the previous chapter, words like *must* indicate irrational beliefs. In this case, the counselor interpreted the client's obsession with perfect grades as an irrational belief.

Although counselors may help clients improve their self-esteem and self-confidence, they should avoid trying to talk clients into improvement. Particularly with persons whose problems are long-standing, verbal persuasion by itself is one of the weakest methods of promoting change (Bandura, 1977, 1997).

CLIENT: For years I've been afraid people will laugh at me if I speak in class.

COUNSELOR: You just have to believe people are going to like you.
  (ill-advised)

CLIENT: I know that, but I can't do it.

Instead, counselors challenge clients to provide proof that their catastrophic beliefs will come true. Frequently, counselors work with clients to design homework to test the rationality of their beliefs.

CLIENT: I'm afraid people will laugh at me if I speak in class.

COUNSELOR: What could you do to test that belief?

CLIENT: Well . . . I could ask a question.

COUNSELOR: That makes sense to me. How about asking one question in each of your classes and seeing how people respond?

At the following session, counselor and client would discuss the results of the homework. In the preceding example, the client would likely learn that no one laughed at her questions. Even if a classmate later ridiculed the client, the counselor might acknowledge the unpleasantness of that event but point out that the client did survive it without any real catastrophic outcome.

26 MAKE PSYCHOLOGICAL ASSESSMENTS, NOT MORAL JUDGMENTS

One of the most difficult tendencies to alter, in some beginners, is judging people. Being judgmental involves moral or ethical assessments.

CLIENT: So I began to drink again after I lost my job.

COUNSELOR: You really let your family down there, didn't you?

Instead of judging whether behavior is right or wrong, counselors should assess clients in terms of psychological theory and practice. Such assessments might involve inquiry into family background, educational and employment experience, psychopathology, intellectual abilities, physical health status, and situational influences (see Lichtenberg & Goodyear, 1999).

CLIENT: So I began to drink again after I lost my job. I sat at home a lot, and I just kept getting more irritable.

COUNSELOR: It sounds as if you felt very stressed and then started to drink again.

In this example, the counselor suggests a psychological cause for the client's drinking, a cause that the client can influence and for which the client bears responsibility. No condemnation of the person is involved.

Sin is a religious term, not a counseling concept. Many clergy perform pastoral counseling in which they serve not as preachers but as skilled listeners and helpers. Some counselors assess the role of spirituality in their clients' lives and discuss those beliefs, at a time the counselor judges appropriate, during the counseling process. Some beginning counselors, however, view counseling as a means of promoting religious values to solve clients' problems. Although it has well-intentioned and skilled followers, religious counseling is a contradiction in terms if practitioners intend, through their counseling, to save or convert their clients. Clients have a right to their personal values.

27 DO NOT ASSUME THAT YOU KNOW CLIENTS' FEELINGS, THOUGHTS, AND BEHAVIORS

When we talk with other people, ordinarily we assume that we know what they are feeling or thinking. Experience in counseling demonstrates the fallacy of this assumption.

> **CLIENT:** So I just left after we started to argue.
>
> **COUNSELOR:** You were very angry.
>
> **CLIENT:** No, I had to leave for work in 10 minutes, and I knew we'd never resolve anything in that amount of time.

In this example, the client corrected the counselor's misperception. In practice, many clients will tell you when you are wrong. Listen for these corrections. To avoid the feeling that they know everything, counselors proceed with a sense of tentativeness (Egan, 2001). They act as though they might have misunderstood, and they ask their clients for confirmation of what they know. They verify their assumptions about clients' feelings, thoughts, and behaviors.

> **COUNSELOR:** You seem sad as you speak of your father. Is that how you feel?
>
> **CLIENT:** No, I didn't feel sad as much as I felt angry about getting into another argument.

Counselors express their tentativeness in this form: a reflection of feeling or content followed by a question. Even if the counselor's reflection misses, clients will respond with further elaboration.

Tentativeness reminds counselors of their ignorance of clients' subjective worlds, and it also provides room to make mistakes. Beginning counselors sometimes feel they cannot utter a single sentence that could possibly harm a client. Counselors *can* make damaging remarks: A reviewer of this book relayed a story of a student counselor, confronted with a sobbing client, who told the client that she was not ready for counseling and should return when she was less upset. The client, shaken by the experience, went to the emergency room of the student health service. Obviously, a tentative, gentle exploration of the client's feelings and presenting problem would have been a much more appropriate way for the student counselor to proceed. Even this client, however, had the wherewithal to seek out help from another source.

28 DO NOT ASSUME THAT YOU KNOW HOW CLIENTS REACT TO THEIR FEELINGS, THOUGHTS, AND BEHAVIORS

Clients differ in their perceptions and reactions to the events in their lives and even to their psychological states. For example, one client may panic when he begins to feel depressed, believing this to be the first sign of mental illness; another may accept depression simply as a sign of physical and emotional fatigue. Be wary of assuming that how *you* respond to an event or feeling corresponds to how your *client* reacts. Observe how clients react to their psychological states.

Clients sometimes create additional problems by how they perceive the original difficulty.

> **COUNSELOR:** So you feel angry toward your son when he deliberately disobeys you.
>
> **CLIENT:** Yes.
> *(tentatively)*
>
> **COUNSELOR:** I sense you're uncomfortable with that anger.
>
> **CLIENT:** I get scared whenever I get mad at him. Good mothers don't get angry.

In this example, the client was troubled both by her son *and* by her anger toward him. What sometimes frightens clients about their feelings is the belief that they must *act* on every feeling. For example, if parents feel angry toward their children, they may feel like hitting the children. Typically, parents may then reject the behavior (hitting) and the feeling (rage, anger). Clients can learn that it is acceptable to experience any feeling but that one should not act on every feeling. Feelings are indicators of our psychological states, not the sole determinants of our behavior. We have choices.

Other clients may be afraid of or confused by their feelings. For example, persons mourning the death of a loved one may be surprised by the type and intensity of their emotions.

> **CLIENT:** I'm just beginning to realize how much I hated my father. But I always thought I loved him!
>
> **COUNSELOR:** You're starting to feel angry about his abuse of you as a child.
>
> **CLIENT:** I think I'm going crazy!

Reassurance helps with such clients, as does group counseling with others going through similar experiences.

> **CLIENT 1:** This is hard for me to admit . . . but I've had really negative feelings about my father since he died.
>
> **CLIENT 2:** What kind of feelings?
>
> **CLIENT 1:** Well . . . anger, mostly.
>
> **CLIENT 2:** I probably felt most angry with my father right after he died.
>
> **CLIENT 1:** Really?
> *(surprised)*
>
> **CLIENT 2:** Yeah . . . I don't think that's unusual at all. I know a lot of people who had to deal with that anger.

Client 1 learned that sadness is not the only feeling people can experience when someone close dies. In sum, clients' appraisals of thoughts and feelings may result in even more intense problems (such as panic attacks) if those appraisals indicate that the feelings are wrong, are inappropriate, or fall short of imaginary standards. Help clients learn how they react to themselves.

SUMMARY AND DISCUSSION QUESTIONS

Clients may sometimes resist change, but counselors can also get in the way. It is not the counselor's job simply to agree with clients, espouse positive thinking, or judge clients. Counselors learn to recognize that conversations with clients are different from ordinary conversations where we typically assume that we understand what others are feeling or thinking. Counselors are aware of their limitations when it comes to understanding their clients' subjective worlds and work hard to explore thoroughly what clients really mean and feel. Counselors listen well enough to understand how clients' experiences differ from, and are similar to, their own. These acts help deepen counselors' understanding of their clients and put counselors in a better position to facilitate client change.

To help you understand and apply the information in this chapter, consider these questions:

When you began your counseling training program, what mistaken assumptions did you hold?

Provide an example of a more complex process of change for a client (for example, see Teyber, 2005).

How might one's religious beliefs help or hinder effective counseling?

What factors might inhibit a client's ability to make rational choices?

Important Topics

The topics in this chapter range from major new forces in counseling to the seemingly mundane. We address important differences in the counseling process and client characteristics as well as several pragmatic aspects of counseling. For many clients, the counseling process is not a matter of meeting regularly over an extended period; crises often bring clients into counseling, and in these instances change must occur quickly. Although the counseling process is often not uniform, neither are clients. Kiesler (1971) developed the term *uniformity myth* to describe the belief that all clients are basically alike. Clients are different, and in this chapter some potentially important differences, such as gender and race, are discussed. We also outline important contemporary developments, such as the rise of managed care and the Internet. Finally, practical issues such as documenting the counseling progress and the referral process are briefly discussed. These details can have important effects on the counseling process.

29 DEVELOP CRISIS INTERVENTION SKILLS

Because of the stigma sometimes associated with counseling ("Only sick people see counselors"), it may take a crisis to motivate some people to seek therapy. Crises occur when individuals are faced with overwhelming problems that they feel they cannot handle (Caplan, 1961; Jacobs, 1999; Puryear, 1979; Slaikeu, 1990). A person in crisis could be, for example, someone who has been depressed for a long period or someone who is experiencing an acute psychotic episode.

Counselors agree about basic steps for working with crisis clients. These actions, described below, differ from counselors' normal behaviors in the initial stage of counseling. For many beginning counselors, the most frightening type of client in crisis is a person considering suicide (Westefeld et al., 2000). First, follow up on any mention of suicide or indication that clients have thought about harming themselves. Don't worry that you will be giving clients a new idea if you ask about suicide. Clients may feel relieved when asked because suicide is so difficult to discuss with friends and family.

> **CLIENT:** I've been depressed for so long now that I've just about given up hope. I don't see any way out.

> **COUNSELOR:** You seem so hopeless that I wonder if you have thought about hurting yourself.

A client might respond like this:

> **CLIENT:** Oh, no, I'd never do that. But I might have to move the family away from here to look for work.

Some clients, however, have considered suicide. Counselors then determine *lethality*— that is, how likely these clients are to attempt suicide. (For a more complete discussion, see Koocher & Pollin, 1994; Slaikeu, 1990; Sommers-Flanagan & Sommers-Flanagan, 2002.)

> **CLIENT:** Yes . . . I just want to die.

> **COUNSELOR:** Have you thought about a way to kill yourself?

> **CLIENT:** Yes . . .

> **COUNSELOR:** Please tell me about it.

> **CLIENT:** I have an empty gun in my bedroom drawer.

> **COUNSELOR:** You've had serious thoughts about doing this. Have you been thinking about buying bullets?

> **CLIENT:** No, not really. I just feel like I want the gun around.

The first question to ask, then, is, *Does the client have a method in mind?* The more specific, accessible, and concrete the method, the greater the likelihood the client will attempt suicide. Without knowing more than the information contained in the preceding example, we would estimate this client's lethality risk as moderate to high. A method is available, but additional action is necessary before suicide would be imminent.

The second question should elicit information about the existence and nature of previous suicide attempts.

> **COUNSELOR:** You have a gun, but no bullets. Have you tried to commit suicide before?

> **CLIENT:** Yes. I tried to slash my wrists. My wife found me, and she rushed me to the emergency room.

The second question, then, is, *Have you previously attempted suicide?* The client in this example indicated that he had made a previous attempt and that it was a serious attempt. Predicting suicide, like predicting other events in counseling, is problematic. The information about a previous, serious attempt, however, combined with an available method, would increase the assessment of the client's lethality to serious. In such a case, counselors ethically and legally have a responsibility to take action to preserve a client's life. This action ranges from further immediate counseling to the client's voluntary or involuntary hospitalization; confidentiality may be broken if the client's life is in jeopardy.

Beginning counselors should learn how to assess for lethality and know what local procedures are employed with suicidal clients. Beginning counselors should *always* consult with supervisors if suicide is a possibility.

Even when suicide is unlikely, counselors working with clients in crisis take more action than they normally do. Such directiveness is embodied in the following overview of crisis intervention strategies.

A. TAKE CONTROL OF THE SITUATION

Particularly with suicidal clients, counselors must take direct action. In most cases, clients have given up control or perceive a great loss of control. The counselor should create structure that promotes order and predictability. In other words, the counselor takes the lead.

> **CLIENT:** Everyone here hates me! I won't leave my room. I'm not going to eat or **(angrily)** sleep or do anything but JUST SIT HERE!

> **COUNSELOR 1:** Jim, I understand you feel you've come to the end of your rope. But you've been sitting on your bed for two days and we can't leave until we can find a way to help you.

> **CLIENT:** No one else will talk to me. Why should you?

> **COUNSELOR 2:** I care about what happens to you. Manuel and I are from the community mental health center and we'd like to listen to what's bothering you . . . .

It's worth noting that, in this example, two counselors worked as a crisis intervention team. Teams of counselors (or of counselors combined with other professionals, such as nurses and police officers) sometimes work as crisis intervenors. A team effort decreases the stress on individual counselors, increases the potential resources available to help the client, and provides greater safety for client and counselors.

B. DETERMINE THE REAL CLIENT

In crisis situations, the identified client may not be the real client—that is, the person who needs attention. Each crisis situation always contains an identified client who, for example, may be threatening suicide or homicide. Identified clients, however, may be calm and clear about what they want. On the other hand, persons dealing with the identified client may be quite anxious and uncertain.

Puryear (1979) describes a crisis he encountered as a counselor in the army. He received a 2 A.M. phone call at home from the emergency room nurse, who anxiously informed him that the emergency room doctor wanted him to come to the hospital immediately. At first, the nurse would not tell Puryear why he should come to the hospital other than to say, "Dr. Smith wants you to come in now." Finally, the nurse whispered that a patient had a gun. Puryear asked that the patient be put on the telephone, learned that the patient wanted to be admitted to the psychiatric ward, and agreed to authorize his admission if the patient gave the gun to the doctor. The patient gave up the gun and Puryear authorized the admission. That resolved the crisis.

In this example, the emergency room doctor and nurse clearly were under a great deal of stress. Nurses, medical doctors, police officers, and other helping professionals with limited mental health experience sometimes become apprehensive when dealing with a crisis client. Your job as a crisis counselor, then, is to assess *all* the persons involved in the crisis situation. Treat everyone with respect, but don't accept information and commands at face value. Work with everyone who is in crisis; practically, this may mean that you will counsel the helping professional who has requested your assistance. Stay in control and avoid getting caught up in others' feelings of panic.

c. EMPHASIZE STRENGTHS

Although many contemporary counseling approaches note the importance of focusing on client strengths (E. J. Smith, 2006), emphasizing strengths is a particularly important method for helping clients regain control in a crisis. (See Puryear, 1979, pp. 86–104, for a more complete discussion.) Even a relatively minor strength or act can be noted.

> **CLIENT:** What can I do? I'm a jerk. I'm failing four classes and nobody will talk to me.
>
> **COUNSELOR:** Hmmm.
>
> **CLIENT:** I can't do anything!
>
> **COUNSELOR:** Well, you can sit in one place for two days and not eat anything. That's a pretty difficult thing to do.
>
> **CLIENT:** Uhmm . . . that's true.
>
> **COUNSELOR:** Maybe you could use that toughness in some other ways.
>
> **CLIENT:** I don't know.
>
> **COUNSELOR:** Well, let's keep talking. But remember that toughness. It could be a big help.
>
> **CLIENT:** Okay.

Note the extent to which clients accept your positive attributions about them. Crisis counselors take control of crisis situations, but they give that control back at a rate consonant with the client's emotional state.

d. MOBILIZE SOCIAL RESOURCES

Find and mobilize clients' social support networks. Determine whether friends and family are available to stay with crisis clients.

> **COUNSELOR:** Bill, I haven't heard you mention your family in any of this.
>
> **CLIENT:** My parents don't care and neither do any of my brothers or sisters . . . except Keisha. And she's in Colorado.
>
> **COUNSELOR:** Tell me about Keisha.
>
> **CLIENT:** She's the person in the family I talk with the most. She convinced me to come to college because I did well in high school.

COUNSELOR: Don't you think she might be concerned about you now, given how you feel?

CLIENT: I don't know.

COUNSELOR: What would you do if she felt depressed?

CLIENT: I'd go visit her . . . maybe I could call her.

Arranging for a crisis client to stay with family or friends provides emotional stability. Once an equilibrium has been reestablished, discuss ways the client could seek further help. If possible, schedule an appointment with a counselor.

Crisis intervention provides important experience for beginning counselors. You will quickly build confidence and knowledge. Client change occurs more rapidly than it does in normal circumstances, and the difficulty level and content of crisis counseling varies greatly. At the same time, continued crisis work over an extended period can be stressful and draining to counselors (see Freudenberger, 1974; Jenkins & Maslach, 1994; Maslach & Leiter, 1997). As a preventive act, many agencies rotate counselors in and out of crisis work to provide periods of less intensive effort.

30 BECOME CULTURALLY COMPETENT

In counseling, culture refers to a community that provides members of the group with information about everything from their worldview to their personal identity and social behaviors. Using this description, culture can include a person's gender, ethnicity, nationality, sexual orientation, occupation, and religion. Given the power of culture, it behooves counselors, as needed with a particular client or group of clients, to become culturally competent. Counselors typically describe this competence in terms of three factors: (a) awareness of our attitudes and possible biases toward individuals of different cultures, (b) understanding a client's worldview in an empathic, nonjudgmental manner, and (c) employing culturally appropriate strategies in working with diverse clients (Engels, 2004). As with counseling in general, being culturally competent can involve learning what *not* to say or do that would impede the counseling process.

Counselors strive to value equality between men and women while trying to understand how sources of inequality affect them (Hill & Ballou, 1998; Kees, 2005; Richardson & Johnson, 1984). Although women make up the majority of clients, attention turned to gender issues in counseling only in the 1970s. Several research studies (see, for example, Broverman, Broverman, Clarkson, Rosenkrantz, & Vogel, 1970; Swenson & Ragucci, 1984; but also see Lopez, 1989, for a different perspective) have suggested that therapists tend to see positive qualities that are most often associated with women (such as valuing relationships) as evidence of mental illness (for example, dependency). At a minimum, female clients should receive nonsexist counseling; that is, they should receive the same type of counseling as males, at least in terms of avoiding bias (Helms, 1979). Specifically, counselors should avoid fostering traditional sex roles, devaluing women, or responding to women as sex objects (compare to Fouad & Chan, 1999; Paniagua, 2005). At the same time, counselors should be familiar with issues that specifically impact female clients (Brown, 1994; Lijtmaer, 1998; Worell & Remer, 2002). Childhood sexual abuse, stranger and acquaintance rape, pregnancy and abortion, and eating disorders are but a few of the issues that occur more frequently for women than for men. In contrast, a concern like restricted emotionality occurs more frequently with male clients.

Counselors also need to be aware of the effects of sexism in our culture on clients (Lott & Rocchio, 1998; Richardson & Johnson, 1984). Sexism refers to discrimination on the basis of gender. Examples include sexual harassment and violence, media representations of women as sex objects, and sex-role stereotyping of occupations (S. Abramowitz et al., 1975). Counselors should not make the mistake of attributing genuine environmental obstacles to clients' internal dysfunction (that is, problems within the client).

> **CLIENT:** I'm having a lot of trouble finding a job. Before my husband and I divorced, I had been a wife and mother for 15 years. No one wants to hire an older woman with no skills.

> **COUNSELOR:** It's really tough to find a job that takes advantage of the important skills you developed during your marriage. I know a program designed specifically to help women in your position find employment. I think it can help.

Like gender issues, racial and ethnic issues have been relatively neglected in counseling and psychology until recently (Carter, 1991; Helms & Cook, 1999; S. Sue, Zane, & Young, 1994). Counselors now strive to understand clients' behavior in the context of their cultural background (Fischer, Jome, & Atkinson, 1998; Montague, 1996; Ramirez, 1999; D. W. Sue, 1990; D. W. Sue, Ivey, & Pedersen, 1996; D. W. Sue & Sue, 2002). Counselors should take responsibility for learning about the cultures of clients different from themselves so that they can best meet their clients' needs. For example, Juang and Tucker (1991) found differences in the amount of verbal communication between Taiwanese and Caucasian American married couples; these researchers also found a significant positive correlation between nonverbal communication and marital adjustment in the Taiwanese, but not the Caucasian, couples. Counselors are wise to attend to the strengths in clients of color as well as to the difficulties resulting from living in a Eurocentric culture. And in instances where the counselor is unfamiliar with a client's culture, the counselor should seek out appropriate supervision.

Counselors should pay attention to their stereotypes and biases about people who are different from themselves. Such misconceptions can inappropriately change the counseling process. For example, you might alter your idea of what constitutes healthy behavior on the basis of race, or you might change your judgment about the appropriateness of intensely expressed emotion on the basis of gender.

Although research findings are controversial (see Casas, 1984; S. Sue et al., 1994), some agencies match counselors and clients by race, hoping that this assignment will facilitate counseling progress (Flaskerud, 1990; B. Pope, 1979). The assumption is that the greater the similarity between counselor and client (in terms of race, sex, culture, and so on), the more easily a therapeutic relationship will develop. Research indicates that in general, clients of color prefer counselors of the same race/ethnicity; in some studies, lower dropout rates and greater attendance have been shown to occur when counselors and clients are matched by race (Fischer et al., 1998). However, Coleman, Wampold, and Casali (1995) view such results as indicating that clients of color see counselor race/ethnicity as a signal of shared attitudes and worldview, thereby enhancing counseling process and outcomes. Atkinson and colleagues (see Fischer et al., 1998) performed a series of studies that found that minority participants strongly preferred counselors with similar attitudes and values—even when given the choice of counselors with similar ethnicity, personality, education, age, and sex. Perceptions of a shared worldview, then, appear to be important, and consequently it can be useful to explore, early in the counseling relationship, clients' possible discomfort when working with a counselor of a different race/ethnicity.

Finally, counselors should not assume that a client's sexual orientation is heterosexual. Whereas some gay and lesbian clients are open about their sexuality, some may not feel safe sharing this with you. Others are confused and still struggling to identify or accept their sexual orientation (see Clunis & Green, 2005).

Successful counseling with gay and lesbian clients partially depends upon the counselor's awareness of his or her feelings about and comfort with sexuality. A counselor who considers lesbians or gay men "sick" or "perverted" must work through these biases before attempting to counsel this group of clients (Betz & Fitzgerald, 1993). Counselors must also be sensitive to the social, legal, and economic barriers gays and lesbians face in our society. These obstacles present special difficulties that others in our society do not face.

> **CLIENT:** I'm really in love with my partner and I would like her to meet my family, but I'm afraid of what they will think about me.
>
> **COUNSELOR:** You're worried that they won't accept you if you tell them you're a lesbian.
>
> **CLIENT:** That's right. My father makes nasty remarks whenever the subject of gays comes up. He thinks it's immoral. But I'm tired of lying to them.
>
> **COUNSELOR:** This is a very important issue for you. Let's explore it some more.

One culture that counselors might not typically conceptualize as such is the military. A substantial number of members of the armed forces (and their families) returning from the wars in Iraq and Afghanistan are expected to seek out mental health services for counseling related to grief, loss, and trauma. These individuals, however, live in a culture very different from the civilian population. Wertsch (in Hall, 2008, p. xiii) termed this culture "the Fortress" and used such words and phrases as "authoritarianism, mobility, officer-enlisted class difference, and the all-encompassing warrior mission of continual preparation for war" to describe military life. The concept of honor also holds a central place, whereby the soldier is willing to act for the common good, including giving up her or his life to perform the mission.

In terms of the military culture's possible impact on the counseling process, Hall (2008) noted that a stigma often exists against seeking help from civilian counselors. Some surveys of military personnel indicate that of those who suffer from serious mental health problems, only half were interested in help and only a third actually obtained services (Hall, 2008). Reflecting concerns about how participating in counseling may affect their careers, Hall suggested that counselors explicitly address confidentiality with soldiers and their families and point out that it is courageous to seek mental health services. In addition, Marshall (2006) maintained that "the whole culture of the military is that you don't talk about feelings or emotions" (p. 32). Thus, a typical question such as "How do you feel about that?" may at least initially be unacceptable to the military client.

Finally, counseling is increasingly taking an international focus as globalization of business, education, technology, and communication expands. As professionals who can function as intermediaries between systems and individuals, counselors are often uniquely positioned to help individuals cope with significant changes in culture, school, and employment. To prepare for their roles in international situations, counselors should consider (a) moving from a national multicultural perspective to an international vision (F. T. L. Leong & Blustein, 2000); (b) conducting and disseminating research on the extent to which Western counseling and psychotherapy theories do and do not apply across cultures (Pedersen & Leong, 1997); and (c) increasing collaboration and sharing of knowledge among researchers and clinicians in different countries, including having U.S. students and faculty read counseling literature from other cultures as well as exchange visits between countries (R. T. L. Leong & Ponterotto, 2003; Leung, 2003). Research is essentially just beginning on such topics as the process of introducing mental health services into traditional cultures that are unfamiliar with counseling (Erhard & Erhard-Weiss, 2007) and the mental health needs of immigrants. Some counseling-training programs are also establishing international programs outside the United States, although the balance in these programs between their educational outcomes for international students and their profitability for the home institution has been a source of controversy.

31 BE OPEN TO GROUP AND FAMILY APPROACHES

Many counselors see one-to-one counseling as the treatment of choice for all clients. Clinical experience and some research, however, suggest that substantial advantages exist for group and family approaches (see Bednar & Kaul, 1994; Gurman & Kniskern, 1978; Yalom & Leszcz, 2005). Particularly from an administrative view, groups are very efficient. If one counselor can work with 10 clients in one 90-minute session, that represents considerable savings in comparison to one counselor working ten 1-hour sessions. From a client's perspective, groups provide emotional support, models of coping behavior, and evidence that one's problems are shared by others. For elaboration of these and other advantages of groups, see Corey (1995); Fuhriman and Burlingame (1994); Schneider Corey and Corey (2005); and Yalom and Leszcz (2005).

Selecting clients by similar age or social maturity can promote a homogeneous group that grows in cohesion and helpfulness (Shertzer & Stone, 1980). Members of helpful groups feel part of the group as a whole and benefit from the open giving and receiving of feelings and ideas (see Barrett-Lennard, 1974; Schneider Corey & Corey, 2005).

CLIENT 1: I want to be an engineer, but I really don't know if I can do it.

CLIENT 2: I felt the same way last year. People thought I was odd because I didn't want to be a secretary.

CLIENT 1: That's right. That's what my parents say.

Using criteria such as those reported by Yalom and Leszcz (2005), counselors typically select clients for groups through individual interviews. In this way, counselors can determine if clients possess sufficient social skills and emotional stability to benefit from a group. Other topics of importance to group counselors include the amount of structure and how to handle anxiety and ambiguity in the group (Kaul & Bednar, 1994). Structured groups, for example, tend to focus on psychoeducational tasks (for example, assertiveness training) and run for a limited period (for example, 12 sessions).

In family therapy, counselors often employ a systems approach (see, for example, Ng, 1999). Family therapists see individuals' problems as relating to the social systems to which they belong, the most important of which is the family (deShazer, 1982; Minuchin, 1974). Families often act to preserve their status quo; in some families, this means that members act to keep certain individuals "sick." The first task of the family counselor, then, is to observe the communication patterns of family members (see Satir, 1988). Next, the counselor points out maladaptive interactions and helps the family members change so that a new structure of communication and interaction takes hold.

32 LEARN ABOUT GRIEF, LOSS, AND TRAUMA

With many clients, the counseling process will focus on or include grief and loss (DeSpelder & Strickland, 2004; Rando, 1993; Worden, 2008). Universal and normal, grief-related issues may be a client's presenting problem or appear during the course of therapy. In addition, grief, loss, and trauma are likely to become increasingly frequent topics in counseling over the next 5 to 10 years as soldiers return from the wars in Iraq and Afghanistan. The Veterans Administration, for example, has estimated that 10 percent to 30 percent of the hundreds of thousands of returning military have posttraumatic stress syndrome (PTSD) and will seek mental health care (Hall, 2008). Some estimates indicate that one-third of the U.S. population is affected by what happens to military families (Rotter & Boveja, 1999).

Often the first step in counseling is to assess how a particular client copes with grief, loss, or trauma. Most people, including some mental health professionals, lack a basic understanding of these phenomena and how they can impact the grieving individual as well as their families, friends, and coworkers (T. Frantz & J. Lawrence, personal communication, October 21, 2008). This can result in a fear on the part of the client of exploring and discussing grief and loss issues. Many individuals' reaction to grief is to want the feelings to go away; in practical terms, this means that most people have not thought very deeply about grief, loss, and trauma. In the University at Buffalo grief-counseling course, for example, at the beginning of the semester students are asked if they have had any notable losses in their life. Most will say no; by the end of the semester, however, almost everyone can report important losses they had not previously conceptualized as such (T. Frantz & J. Lawrence, personal communication, October 21, 2008).

Many of the basic listening and counseling skills described in this text can be applied to grief and trauma counseling. For example, one of the counselor's key tasks is to pay attention to the client's experienced and expressed levels of affect (Pennebaker, Zech, & Rimé, 2001). Because individuals, families, and cultures differ in their comfort with the expression of affect, particularly with intense feelings such as those associated with loss and trauma, a conflict can result between the counselor's wish to focus on feelings and the client's reluctance to do so. With many clients, grief counseling involves a balance between accepting the individual as she or he is (that is, pacing the client) and helping the individual to process affect (leading the client). This balance requires training, practice, and patience so that the grief counselor can develop a sense of timing about when to stay with the client's current level of consciousness and when to move toward affect. A client with a terminal illness, for example, might initially be able to talk about the frustrations and pain of dealing with treatment side effects, but not about a fear of dying.

> **COUNSELOR:** We've talked a lot about how you can adjust to the side effects of your chemotherapy . . . but we haven't spoken at all about dying.
>
> **CLIENT:** (*silent, pensive*)
>
> **COUNSELOR:** (*silent*)
>
> **CLIENT:** I know . . . it's just whenever I even think for a second about dying, I just feel terrified.

Unfortunately, recent reviews of research examining the effectiveness of grief counseling have found only small to moderate effects for the average grief-counseling intervention (Allumbaugh & Hoyt, 1999; Jordan & Niemeyer, 2003). Allumbaugh and Hoyt's meta-analysis, for example, found that only self-selected grief-counseling clients evidenced large gains (as compared to clients recruited by researchers). Jordan and Niemeyer suggested that these findings might be explained by three factors: (a) Grief is normally self-limiting, that is, most mourners can work through losses with the help of family and friends; (b) the type of brief grief counseling delivered in research studies is too weak to be effective; and (c) methodological problems, such as inappropriate measures used to measure outcome, may have prevented researchers from detecting larger counseling effects.

In addition, some evidence indicates that roughly one-third of grief-counseling participants worsen over the course of counseling (Fortner, 1999). The benefits and risks of grief counseling are likely to result from an interaction between client needs and the degree and sequencing of affective focus and meaning making. Hall (2008), for example, talked about counseling a group of high school students who had seen a classmate killed in a freak accident; these students had increased absences, lost interest in school work, and evidenced negative changes in behavior and attitudes. Hall reported that she tried to "help them change the focus of their interactions from reliving the tragic event by describing it over and over to each other in graphic detail, which only continued to retraumatize them, to talking about the impact the experience had on them personally" (p. 205). Thus, it appears that grief counselors should frequently assess how clients are experiencing the counseling process as well as the impact of the process on the individual client (for example, are they feeling more anxious and depressed, and are these feelings persisting over time?). For many counselors, this task may be easier to perform in an individual counseling setting.

In contrast, researchers have found that prolonged exposure therapy is effective for working with clients with PTSD (Foa, Hembree, & Rothbaum, 2007). This treatment aims to help clients emotionally process their traumatic experiences so that PTSD symptoms decrease (Foa et al., 2007). In this approach clients (a) learn about common reactions to trauma, (b) learn to relax through breathing techniques, (c) experience repeated in vivo exposure to situations the client has been avoiding, and (d) experience repeated exposure to traumatic memories. This approach is based on learning theory that suggests that individuals may learn to inappropriately associate certain stimuli with feared events and situations. A person who has a serious automobile accident, for example, may come to associate riding in any car with the sights, sounds, and feelings of the accident event. The emotional exposure of therapy aims to show the individual that those stimuli are not always associated with a car accident. To do so, the client experiences the fear and anxiety associated with the accident, repeatedly, while learning that riding in a car does not always lead to an accident.

Another approach with research support, which may be particularly useful for men who are reluctant to engage in personal counseling, focuses on writing about personal traumas (Pennebaker, Mehl, & Niederhoffer, 2003). Participants are instructed to "Write about your deepest thoughts and feelings about a trauma" over a period of three to five days for 15 to 30 minutes a day (Pennebaker et al., 2003); in essence, participants must create a written narrative of a traumatic incident. Research indicates that the writing procedure induces affect and meaning making in ways similar to traditional counseling. Smyth's (1998) meta-analysis of 13 studies comparing the effects of the Pennebaker writing intervention with some type of control (for example, another writing condition, such as "Write about your plans for the day") found that written emotional expression groups exceeded controls on a wide range of outcome measures, including measures of affect, grade point average, immune system performance, number of health center visits, and reemployment status. These studies also suggested greater benefits for males (perhaps because men's sex role inhibits emotional disclosure to other people) and for writing periods that are distributed over a longer period of time. Interestingly, many writers reported an increase in short-term distress before later improvements became evident. Pennebaker et al. believe that "the construction of a story or narrative concerning an emotional upheaval may be essential to coping" (p. 568) and that successful therapy requires individuals to "move from highly specific referential activity and high emotional tone to high levels of abstraction" (p. 568).

33 REFER CAREFULLY

You cannot help every client. You may lack the necessary skills; clients may move to a distant locale; you and a particular client may simply be unable to work together. Regardless of the reason, you should be aware of other resources in the community.

>**CLIENT:** We're moving to Seattle in a month, but I'd still like to continue therapy with someone.

>**COUNSELOR:** Actually, I know several counselors in that area who work as private practitioners. I can give you their names and phone numbers if you like.

>**CLIENT:** That would help. Would you be willing to speak to my new counselor if that's necessary?

>**COUNSELOR:** Yes. You should sign a release of information form so I can do that.

To respect clients' rights and to ensure that they follow through, explain the reason when you make a referral. After your explanation, seek a reaction to your referral.

> **COUNSELOR:** I think a psychological test might save time and help me get a better sense of what is happening with you.
>
> **CLIENT:** What kind of test?
>
> **COUNSELOR:** It's called the Minnesota Multiphasic Personality Inventory, or MMPI-2. It takes about an hour and a half, and it will help us decide what direction to go in counseling. What do you think?
>
> **CLIENT:** That's okay. I think maybe I took that test a few years ago at the school I transferred from.
>
> **COUNSELOR:** Okay. We have to set up an appointment for you at the testing center. Maybe we can also track down those previous results—with your permission.

Moxley (1989) employs the term *linkage* to describe strategies that are employed by mental health professionals to connect clients with needed services through the referral process. To ensure a connection between a client and a new agency, for example, the counselor must first address any fears or misconceptions the client holds about the agency. The referring counselor should also attempt, with the client's written permission, to transmit essential information about the client to the agency. Such information might include the purpose of the referral, specific needs of the client, and relevant assessment data. Moxley also suggests that counselors telephone clients after the referral to be certain that the client did connect with the agency and is receiving adequate services.

34 WATCH FOR DETERIORATION IN CLIENTS

Counselors rarely consider that clients may be harmed by the counseling process (Kendall, Kipnis, & Otto-Salaj, 1992). But considerable evidence exists to suggest that some clients, particularly those with severe problems, deteriorate in counseling (Lambert & Bergin, 1994; Mohr, 1995). In research on groups, for example, Lieberman, Yalom, and Miles (1973; compare to Beutler, 1984) found that leaders who were frequently aggressive, authoritarian, and inappropriately self-disclosing produced the greatest number of dropouts and negative changers (people who became worse after counseling).

No generic prescriptions can be given for avoiding negative change in counseling; however, you should assess and observe client reaction to every intervention. The following interaction, for example, might occur in the beginning stages of counseling.

> **CLIENT:** I have faint recollections of my father hurting me when I was very young . . .
>
> **COUNSELOR:** Did your father abuse you?
>
> **CLIENT:** I . . . I don't know. I'm starting to get very upset.

COUNSELOR: You seem anxious . . .

CLIENT: Yes. I'm very scared.

COUNSELOR: How do you feel about talking about this?

CLIENT: I'm not ready. Not yet . . . at least not this fast.

COUNSELOR: That's okay. We'll come back to this later.

In this example, the counselor assessed the client's feelings about further probing of the potential abuse. Because the counseling process was still in the early stages, the counselor decided to table the probe and reapproach the issue later in counseling.

Pushing clients too fast may be harmful. Some counseling approaches, such as Gestalt, emphasize frustration and confrontation of clients. Gestalt counselors, however, still observe their clients to determine the effects of confrontation.

35 ESTABLISH AN INTEREST IN COUNSELING RESEARCH

Counseling is an art in the throes of becoming a science. At present, practitioners can do more than researchers can explain. For example, counselors can help clients with phobias and depression, but researchers can only partially explain the etiology of those problems and why counseling works with them when it does. A substantial split separates practitioners and researchers: Researchers dismiss practitioners as "touchy-feely," whereas practitioners view current research as irrelevant (Gelso, 1979; Meehl, 1956). However, practitioner-scientists, theoreticians, and other researchers are the groups likely to discover the patterns and order that occur in effective counseling and psychotherapy.

Competent counselors and psychotherapists should understand the research process, if not contribute to it (see Heppner, Kivlighan, & Wampold, 1999). Keep up with current research by reading journals such as these:

American Journal of Psychiatry

Behavior Therapy

Clinical Psychology Review

Cognitive Therapy and Research

Group Dynamics

Journal for Specialists in Group Work

Journal of Consulting and Clinical Psychology

Journal of Counseling and Development

Journal of Counseling Psychology

Journal of Multicultural Counseling and Development

Professional Psychology: Research and Practice

Psychology of Women Quarterly

Psychotherapy Research

Counselors who understand the research process are better able to conduct, cooperate in, and coordinate research in applied settings (such as hospitals, clinics, college counseling centers, and community mental health centers). Applied settings, because of their primary orientation to service, typically present counseling researchers with substantial obstacles to establishing and maintaining research, be it single-case designs or large outcome studies. Conducting research in field settings such as community mental health agencies, however, provides an excellent source of research questions and offers great potential for producing results relevant to actual clients and problems.

Beginning counselors who desire to learn about research should ally themselves with a faculty member or counselor who is conducting empirical work (Gelso, 1979; J. Kahn & Scott, 1998). Such an alliance teaches the beginner research skills and demonstrates the day-to-day difficulties of conducting counseling research. Researchers in applied settings must learn, for example, how to protect their scarce research time against service demands and how to handle the politics of conducting research in a service agency (Gelso, 1979).

36 DOCUMENT YOUR WORK

After each counseling session, take notes (Prieto & Scheel, 2002; Wiger, 2005). Documentation has five benefits:

1. Records provide support for you in any legal action.
2. Records allow you to accurately inform your supervisor of process and issues.
3. Records enable you to comply with agency standards for accountability (such as the number of hours spent in individual counseling or issues to be addressed should a client terminate and later return for further counseling).
4. Records remind you of treatment history and amount of progress.
5. Records provide the basis for financial reimbursement by third-party payers (for example, insurance companies) and for charging fees to clients.

What should you write in your notes? This depends partially on where you work, but in general you should briefly describe for each session (a) principal content and themes; (b) noteworthy counselor interventions, with client response; (c) client status (including descriptions of client cognitions, affect, and behavior); (d) expected client behavior during the period leading up to the next session, including any homework; (e) expected time and interventions still required to meet counseling goals; and (f) indications of client expectations of the outcome of counseling. Your notes could also include tests administered (with results and client reactions), diagnoses, your reaction to the session, questions to ask supervisors or colleagues, and relevant client interactions with family and friends. Such notes should be kept under lock and key where no one can access them purposefully or by accident. Computerized record-keeping systems are available and also must have restricted access.

37 PERSEVERE WITH CLIENTS WHO NO-SHOW

Beginning counselors sometimes have difficulty knowing how to react when their clients fail to attend scheduled sessions. One student counselor had his first three clients fail to show up, and the student assumed that something was wrong with his initial phone contact. Another counselor became angry with a client for failing to attend a session; he subsequently decided not to attend the next session, rationalizing that he wasn't certain the client was going to attend. The client did.

Avoid judgments on clients who no-show. At the next contact (a follow-up phone call or scheduled session), ask the client about the absence. Determine if the failure to attend reflects on progress (or lack thereof) in counseling or the client's characteristic manner of dealing with people. Also determine if the client conceives of time in different ways than you do. If helpful, provide a written reminder, such as an appointment card and a rescheduling method at the end of each session. With special-needs clients, build in some flexibility. In essence, regard failure to attend as material for therapy rather than as a personal affront.

> **COUNSELOR:** Amy, you missed the last two sessions.
>
> **CLIENT:** Yeah. I just didn't feel like coming.
>
> **COUNSELOR:** Tell me more.
>
> **CLIENT:** After our last session I felt so depressed I didn't want to talk about my problems anymore.

Failure to attend can become especially complicated when clients interpret it as an additional failure. For example, clients may miss a meeting after a particularly difficult session; however, they may then see the no-show as an indication that they are unable or unworthy to continue counseling. Premature termination then becomes a real possibility (Kane & Tryon, 1988). You might avoid this failure loop by giving such clients excused absences—that is, planned permission to miss a session. This may increase some clients' willingness to attend future meetings following a no-show.

> **CLIENT:** I missed the first session because I didn't feel like talking. And then I got embarrassed because I didn't show up.
>
> **COUNSELOR:** So you missed the second session because of your embarrassment. Why don't we just agree it's okay for you to miss a session if you call ahead and cancel?
>
> **CLIENT:** Gee . . . that sounds okay.

If *you* must cancel a session, inform the client as soon as possible and reschedule. Also, many agencies and individual practitioners charge a fee for missed appointments; clients should be informed in advance of such policies.

38 LEARN HOW TO CONCEPTUALIZE CLIENTS

Creating a visual model of the counseling process and outcome can be a useful method for understanding the complexities of working with clients (Eells, 1997; Meier, 1999, 2003). Models are graphical representations of important elements influencing the behavior of a particular client. All counselors develop ideas about what is troubling their clients and what they could be doing to help them (Eells, 1997; Haynes, Leisen, & Blaine, 1997; Meier, 2003; Spengler, Strohmer, Dixon, & Shivy, 1995; Teyber, 2005). These mental models, however, are often implicit and, occasionally, ineffective. For example, new counselors often act as if the best way to intervene with clients is to give advice. This model—that advice translates into problem resolution—is typically an inaccurate depiction of the needed counseling process.

One source of information for case conceptualizations is empirically supported treatments (that is, counseling approaches based on research evidence and described in treatment manuals). Counseling and psychotherapy researchers have conceptualized and evaluated treatment approaches that may be applicable to many clients who present with common problems such as depression and anxiety. This nomothetic approach is further described in Section 58 in Chapter 6.

More idiographic case conceptualizations can be created through close observation of clients and counselors. Close observation in this context refers to (a) representing our experiences with and ideas about clients in terms of models specific to an individual, and then (b) assessing the efficacy of these models. Suppose, for example, that Robin, a high school student, begins to have frequent fights at school after she learns that her parents are considering a divorce. When Robin gets into a fight, several outcomes occur, as shown here.

Process   *Outcomes*

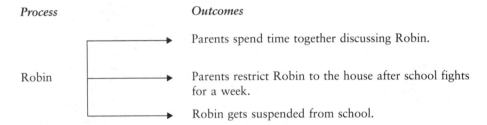

Robin   Parents spend time together discussing Robin.

Parents restrict Robin to the house after school fights for a week.

Robin gets suspended from school.

Note that when Robin gets in trouble at school, several events occur that differ in their attractiveness to Robin. For example, Robin may be ashamed at being suspended and angry about being restricted but secretly pleased that her arguing parents must spend time together to decide what to do about her. Although this is a simplified conception of what may be happening with Robin and her family, it is a useful starting point for considering how best to help Robin.

Given this knowledge, we could represent the causes of Robin's problem as:

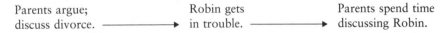

Parents argue;   Robin gets   Parents spend time
discuss divorce.   in trouble.   discussing Robin.

This model represents our initial ideas about the causes, origins, or development of Robin's problem. We can also include in the model the procedures we believe are necessary for the resolution of Robin's problem:

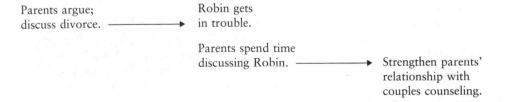

Parents argue;   Robin gets
discuss divorce.   in trouble.

Parents spend time
discussing Robin.   Strengthen parents'
                             relationship with
                             couples counseling.

Thus, our initial guess about how to help Robin involves a family systems approach. That is, we speculate that the source of Robin's problems lies in the family, and not in Robin herself (Bowen, 1978; see also Chapter 6). We suspect that if we strengthen the parents' relationship or the family as a whole, Robin will get something she needs and this will decrease the frequency of her getting in trouble at school.

Again, this conceptualization is just a starting point in a treatment plan. The model may turn out to be completely wrong, very helpful, or, most likely, in need of some modification. One of the interesting aspects about assessment in counseling is that clients often recall new or different information as they delve deeper into self-exploration (compare to Schwarz, 1999). The importance of a model is that it can serve as one source of information about how to proceed in counseling.

The next steps would involve collecting qualitative and perhaps quantitative data that would enable the counselor to chart trends in this client's process and out-come variables. J. S. Abramowitz (2002), for example, reported a case study with a 19-year-old Asian American male whose attempts to control his aversive thoughts were resulting in increased anxiety and avoidance of social situations. The therapist collected idiographic (that is, pertaining to a particular individual) data on three specific measures (that is, the client's fear of intrusive thoughts, avoidance of situations associated with intrusive thoughts, and neutralizing rituals). Abramowitz found improvement on all three measures over 16 sessions; more details about the case conceptualization and idiographic measurement process with this case can be found in Meier (2008). Thus, idiographic data that flow from the elements of a case conceptualization can provide documentation about counseling progress as well as indicate modifications to the model that could result in more effective inter-ventions (Claiborn & Goodyear, 2005; Hartman, 1984; Paul, 1986).

Although case conceptualization, outcome assessment, and treatment planning are important, they are also time-consuming. In some settings, it may be possible to increase the time for these activities by slightly decreasing the amount of client contact time. For example, if you typically spend 50 minutes with an individual client, consider making the session length 45 minutes (starting with new clients). This 15-minute period between sessions can provide the time necessary to take good progress notes as well as review the case conceptualization and assessment trends for the next client.

39 LEARN ABOUT MANAGED CARE

The single most powerful force affecting the profession of counseling since this book's first publication, we believe, has been managed care (Davis & Meier, 2000). Designed to halt significant increases in health care costs, managed care has had three general effects (compare to Kassan, 1996; Kessler, 1998; Shore, 1996):

1. Health care expenses have stopped their rapid growth, at least temporarily.
2. Counselors in private practice now struggle to provide useful services to clients in the context of overwhelming constraints put on them by managed care companies.
3. Managed care methods have been adopted or imposed upon most counseling agencies.

The constraints placed by managed care on counselors and counseling agencies have been systematic and substantial. To control costs, companies have employed a multitude of methods, including limiting the number of sessions, restricting who may provide services, delaying payment for sessions by weeks and months, and continually pressing for or imposing lower fees (Kessler, 1998). Brief therapy is now the norm in most settings; some insurers limit coverage to two to three sessions annually (Shore, 1996). Essentially, the majority of managed care companies have demonstrated by their behavior that they value profit over quality care, in diametric opposition to the values of most counselors and health care professionals (compare to Gray, 1991).

Managed care often means that counselors must significantly adjust the counseling process. In private practice, for example, role induction may need to include an explanation that the number of sessions typically allowed (for example, two to three *total*) does not match the number of sessions the insurer advertises to clients (for example, 20). Counselors may also need to explain (a) the limits of confidentiality (that is, the employees of the managed care company may also view client information) and (b) that the insurer will attempt to influence whether counseling continues or ends as well as which counseling approach should be employed.

In our view, managed care to date has had a negative influence on most counselors and clients. Counselors are frustrated by the micromanagement of managed care, the pressure to conclude counseling prematurely, the medical approach imposed on the counseling process, and the constant introduction of new methods designed to delay or deny services and payments. Counselors must cope with increased paperwork, such as frequent written progress reports required by the insurer to continue payment for treatment. Counselors in private practice no longer may operate as solo practitioners but must set up an office in a group practice, complete with secretary, billing services, and crisis services. Counselors must often pay as much attention to the business aspects of their practice as to the counseling component.

Despite our overall negative assessment, managed care's emphasis on accountability may eventually help to improve counseling. Counselors and clients now must attempt to document the effectiveness of counseling through the use of outcome measures. Outcome measures are tests and rating scales designed to gauge a client's progress. In contrast to the use of idiographic measures tailored to specific clients, typically the field has tried to employ nomothetic measures (that is, general tests intended to be suitable for all clients, such as the MMPI) developed for other purposes (for example, diagnosis) for use as outcome indicators. More contemporary outcome assessments such as the Global Assessment of Functioning Scale (GAF; American Psychiatric Association, 2000) and the Outcome Questionnaire (OQ-45; Wells, Burlingame, Lambert, Hoag, & Hope, 1996) examine a client's symptoms or functioning in areas such as interpersonal relationships, work, and school (Froyd, Lambert, & Froyd, 1996; Ogles, Lambert, & Masters, 1996). Many current measures, however, often are too long for repeated use, have not been evaluated in terms of their ability to detect change resulting from psychosocial interventions, or have other sources of invalidity that raise questions about the extent to which their scores reflect change from counseling versus other sources. Current outcome measures have been shown to provide (a) useful feedback to counselors about the progress of individual clients, particularly in terms of identifying failing clients (Lambert et al., 2001) and (b) information about the effectiveness of counseling at the level of an agency (that is, a program evaluation for a group of counselors and their clients; Meier, 2008). In contrast, idiographic measures and nomothetic measures specifically designed for outcome measurement may help provide more precise and valid feedback to individual counselors about how to adjust the counseling process and intervention to improve client progress (Meier, 1999).

The potential also exists to improve managed care itself. As discussed in *The Elements of Managed Care* (Davis & Meier, 2000), we suggest that counselors do not simply acquiesce to managed care methods but adapt to survive in the current environment and organize to change the worst abuses. Among the ways to pursue the latter goals are to join professional groups that challenge managed care and to support organizations that promote fairness as discussions about health care reform proceed in the United States.

40 DEVELOP TECHNOLOGY SKILLS

Although counselors have long considered the potential uses of technology (for example, Meier, 1986), the rise of the Internet has spawned renewed interest and a greater number of applications (Sampson, Kolodinsky, & Greeno, 1997). Current applications include enhancing the information provided and used during career counseling (Sampson, 1999; Stevens & Lundberg, 1998), using e-mail as part of supervision (Myrick & Sabella, 1995), and supporting discussion and support groups (Bowman & Bowman, 1998).

Providing information for counselors and clients remains the most important current application on the Internet. As noted in Chapter 1, most professional organizations now offer websites with information such as ethical codes. These include:

www.aamft.org	American Association for Marriage and Family Therapy
www.abct.org	Association for Behavioral and Cognitive Therapies
www.apa.org	American Psychological Association
www.asgw.org	Association for Specialists in Group Work
www.counseling.org	American Counseling Association
www.cpa.ca	Canadian Psychological Association
www.nasponline.org	National Association of School Psychologists
www.naswdc.org	National Association of Social Workers
www.nbcc.org	National Board for Certified Counselors

www.ncda.org	National Career Development Association
www.arcaweb.org	American Rehabilitation Counseling Association
www.psych.org	American Psychiatric Association
www.psychologicalscience.org	Association for Psychological Science
www.schoolcounselor.org	American School Counselor Association

Other sites provide general information about mental health topics, specific counseling approaches, and psychology:

www.adaa.org	Anxiety Disorders Association of America
www.beckinstitute.org	The Beck Institute for Cognitive Therapy and Research
www.rebt.org	Albert Ellis Institute (Rational-Emotive Behavior Therapy)
www.psychwatch.com	Practitioners' online resources

Obtaining quality information from such sites, however, can be problematic. Unlike journals and books, websites typically do not benefit from peer review; the result is that the information provided ranges from poor to excellent. Also, some web-sites frequently change URL (universal resource locator) addresses; the addresses listed here may or may not remain valid when you read this book. All of these addresses, as well as related websites (such as those addressing managed care issues), will be updated and can be accessed from the first author's home page:

www.acsu.buffalo.edu/~stmeier.

Efforts to provide counseling over the Web have had mixed results and seem un-likely to become widespread in present forms. Although the Web may make counseling-related services more accessible to some individuals, Web counseling lacks theoretical models of practice, supporting empirical evidence, and methods for guaranteeing confidentiality (Heinlen, Welfel, Richmond, & Rak, 2003). Heinlen et al. recently examined 136 websites that offered counseling primarily through com-puter chat rooms and e-mail. They found widespread lack of compliance for Web counseling standards offered by the National Board for Certified Counselors (NBCC, 1997). For example, one standard requires verification of parent/guardian consent in cases where the client will be a minor; however, only 26 percent of the websites referred to issues involved in the counseling of minors (Heinlen et al., 2003). Heinlen et al. concluded that "of the 136 sites sampled in this study, not a single site was in full compliance with the NBCC Standards" (p. 66). In addition, a follow-up of the websites eight months after the initial examination found that 37 per-cent were no longer operating.

It remains a safe prediction that the Web will continue to grow in importance, for both clients and counselors, as a source of information. What remains to be seen is the extent to which other types of counseling applications will develop and find acceptance on the Web.

SUMMARY AND DISCUSSION QUESTIONS

This chapter makes clear that different methods of counseling exist. Part of the counselor's job is to become skilled in at least a few modalities and with at least a few client populations. Counselors learn the limits of their competencies and how to refer clients they cannot help. The professional counselor also stays aware of other approaches as well as new research, technological, and practical developments in the field.

To help you understand and apply the information in this chapter, consider these questions:

What are the major questions and actions involved in crisis intervention?

How might issues of ethnicity (your ethnicity as well as the client's) impact your counseling?

Assuming you are an average counselor, what percentage of your clients might be expected to NOT make progress? (To answer this question, you might consult with a supervisor, faculty member, or practicing counselor.)

How might a client's age affect the counseling process?

How will you react when a client no-shows?

CHAPTER	5	COUNSELOR, KNOW THYSELF

In no other profession does the personality and behavior of the professional make such a difference as it does in counseling (Kassan, 1996). Beginning counselors need to work at increasing their self-awareness as well as their knowledge of counseling procedures. Your willingness to be open to supervision, to accept clients' failures and criticisms, and to acknowledge your limits will contribute to your eventual success and satisfaction. Some faculty also recommend that students participate in counseling themselves to gain a better understanding of the counseling process as well as increased self-understanding.

41 BECOME AWARE OF AND ADDRESS YOUR PERSONAL ISSUES

Every beginning counselor will eventually confront difficult questions about her or his personal issues. Answers to these questions may not be readily apparent. Your issues, however, do influence how you counsel (see Egan, 2001; M. Kahn, 1997; Kottler, 1997). Research indicates that very few graduate programs require psycho-therapy for their students, although 70 percent of therapists in a recent survey believed that therapy should be required for therapists-in-training (Pope & Tabachnick, 1994). Counseling others and being counseled, having competent supervision, and developing a theory of counseling will help you answer the following questions.

A. HOW DID YOU DECIDE TO BECOME A COUNSELOR?

Many counselors became interested in a helping profession because family and friends sought them out to listen and provide help. Other counselors were once clients themselves and decided to follow in the footsteps of counselors who helped them. Are there aspects of your family, gender, or culture that led you to a counseling profession?

Your reasons for becoming a counselor may affect how you counsel. For example, did your assistance to friends include acting as a rescuer? Rescuers are persons who do the problem solving and helping in their social groups. As a counselor, will you again rescue?

CLIENT: I just don't have the willpower to eat more.

COUNSELOR: You can call me any time of the day or night when you need help.

Or can you allow clients to handle their own problems?

CLIENT: I just don't have as much willpower as I once had.

COUNSELOR: What stops you from using the willpower you do possess?

A counselor can solve a problem for a client. A counselor can also help clients learn how to solve problems successfully. The difference is important.

B. WITH WHAT EMOTIONS ARE YOU UNCOMFORTABLE?

Clients often experience intense feelings such as rage or grief. Because of their lack of exposure, beginning counselors may be uncomfortable with some feelings, at least at the intensity level expressed by clients. Will you allow clients to express those emotions? Are there feelings that you avoid, that you will steer your clients away from?

CLIENT: I just hurt so much since the divorce . . .

COUNSELOR: Try not to think about it. Are you having any luck meeting people?

In this example, the counselor unconsciously moved the discussion away from the intense hurt. Counselor training includes exposure and desensitization to a range and intensity of feelings new to beginning counselors. Identify the issues that make you uncomfortable: For example, as a new counselor, one of the authors was afraid of talking with clients about their sexuality (particularly homosexuality), their feelings of rage, and their feelings of helplessness. Without help from supervisors and the experience of personal counseling, the author would have continued to avoid those areas when clients ventured near them.

C. WHAT AMOUNT OF PROGRESS IS ACCEPTABLE?

Although individuals differ in the progress they make in counseling, beginning counselors are sometimes surprised at how slowly clients change. Most clients cannot alter their troublesome behavior quickly or at the instruction of the counselor. In fact, research suggests that clients with more severe problems will change more slowly and often experience poorer outcomes (Lambert, 2005). For counselors, the issue becomes how much progress is acceptable.

CLIENT: I know I agreed to speak in each of my classes, but . . .

COUNSELOR: *(silence)*

CLIENT: . . . but I asked only one question, in physics.

What will be a blow to your confidence as a counselor? A client who drops out? Ask yourself what rewards you expect as a counselor. Do you want money, status in the community, or intimacy without threat?

D. HOW WILL YOU DEAL WITH YOUR CLIENTS' FEELINGS TOWARD YOU?

If the counseling relationship becomes at all significant to the client, she or he will have feelings toward you. Interestingly, these feelings may have little to do with what you've said or done. Clients may transfer feelings from past relationships onto their perceptions of you. Clients' feelings about their counselors are generally referred to as *transference*. (See Hansen et al., 1994, for a general introduction; see M. Kahn, 1997, and Robertiello & Schoenewolf, 1987, for advanced examples.) How will you feel if a client perceives you as attractive, wise, or racist?

> **CLIENT:** I know you don't like me.

> **COUNSELOR:** What makes you say that?

> **CLIENT:** You don't like me because I'm blind!

In general, counselors process such feelings with their clients. Because this processing may influence progress, it is important to develop your ability to recognize and work with clients' feelings for you.

One area particularly difficult for new counselors to address is sexual attraction between client and counselor. Despite the fact that sexual encounters between counselor and client typically have negative effects, particularly for the client (Pope, 1988), research suggests that about 5 percent of therapists have engaged in such behavior (Rodolfa, Kitzrow, Vohra, & Wilson, 1990). That such sexual intimacies occur is not surprising given that over 80 percent of psychologists surveyed reported feeling sexually attracted to a client (Rodolfa, Yurich, & Reilley, 1993) and that counselor-client sexual interactions are often the focus of ethical complaints and malpractice awards (Rodolfa et al., 1994). Consequently, it is important that counselor trainees learn how to handle the common situation of counselor-client attraction through supervision and training experiences (compare to Rodolfa et al., 1990).

E. HOW WILL YOU HANDLE YOUR FEELINGS FOR YOUR CLIENTS?

You may have strong or ambivalent feelings about clients. Counselors' feelings about their clients are generally referred to as *countertransference*. (See Hansen et al., 1994; M. Kahn, 1997; Robertiello & Schoenewolf, 1987; Silverstein, 1998.) Is it okay for you to feel sexually attracted to a client? What will you do with that feeling? Perhaps you came from a family in which you were abused as a child. Will you be able to work with child abusers? You too may transfer your feelings about past relationships onto your clients.

> **CLIENT:** I've always gotten my way. I was always the biggest kid in grade school, and I just bullied anybody who got in my way.

> **COUNSELOR:** Don't you think you were a jerk to do that? (*angrily*)

Counselors-in-training (and professional counselors, for that matter) often engage in their own therapy both for personal growth and to be better able to help their clients.

> **CLIENT:** I've always gotten my way. I was always the biggest kid in grade school, and I just bullied anybody who got in my way.
>
> **COUNSELOR:** Anyone who gets in your way gets pushed aside.
>
> **CLIENT:** Yeah . . . but bullying doesn't always get me what I want with my family and coworkers.

On the other hand, the relationship between counselor and client sometimes parallels other relationships. For example, if sarcasm is part of a client's interpersonal style, sarcasm is likely to be part of the client's verbal behavior with the counselor. What you as the counselor feel about the client, as a person, may also reflect what other people feel. Some counselors judge progress in counseling (particularly for clients with interpersonal difficulties) by changes in the relationship between counselor and client.

> **COUNSELOR:** You know, John, I sense that you've felt more relaxed talking with me the past couple of weeks.
>
> **CLIENT:** I guess I don't feel quite as suspicious of people as I used to. Funny—a friend of mine said the same thing last week.

In this example, the counselor noticed less tension in the relationship. The client acknowledged a change in his interpersonal style both in and out of counseling.

F. CAN YOU BE FLEXIBLE, ACCEPTING, AND GENTLE?

The attitudes you hold toward counseling and clients affect the process and outcome of counseling. Flexibility refers to the counselor's ability to be creative, open, and aware of the here-and-now in counseling (Hansen et al., 1994; Van Kaam, 1966). Acceptance is the counselor's willingness to hear and understand whatever the client says without judgment, without conditions. Gentleness is the capacity to be kind and considerate even when clients are abrupt, afraid, and defensive. These counselor characteristics aid the development of the therapeutic alliance, an important contributor to positive client outcomes (Lambert & Bergin, 1994).

> **CLIENT:** You're not helping me. I can't talk about this!
>
> **COUNSELOR:** How difficult it is for you to remember your childhood . . .
> **(softly)**
>
> **CLIENT:** It hurts so much.
>
> **COUNSELOR:** *(silence)*

As a profession, counseling has members who are probably more accepting of individual differences than professionals of any other group. Such tolerance can be learned.

42 BE OPEN TO SUPERVISION

The preceding set of questions may be overwhelming to beginners, who may wonder how they will ever become so self-aware. Relax—no counselor is expected to be self-actualized. You are expected, however, to know the areas where you have personal difficulties. Counseling involves knowing yourself, not just knowing techniques and theory; in this area, a competent supervisor can be invaluable. Supervision is more than instruction, consultation, or direction (see Holloway, 1995; Stoltenberg, McNeill, & Delworth, 1998; Whiteley, 1982). Supervision resembles counseling when it involves exploring the supervisee's personal issues.

> **COUNSELOR:** So when he started to talk about dying, I really became anxious. I started to sweat and feel really dizzy.
>
> **SUPERVISOR:** What was it that frightened you?
>
> **COUNSELOR:** I don't know.
>
> **SUPERVISOR:** It was when he mentioned dying that you became anxious.
>
> **COUNSELOR:** I just haven't dealt with death much, and he was feeling so afraid. That triggered the same feelings in me.

Supervision is the best setting in which to explore feelings about your clients. Of course, beginning counselors may find it useful to enter counseling themselves. Supervision cannot be entirely concerned with the supervisee's issues, and personal counseling may facilitate the new counselor's growth as a therapist. You also gain the perspective of seeing what it's like to be on the other side of the counselor-client relationship. We consider supervision so important that we strongly recommend that you continue supervision after your official training ends. Maintaining the discipline and perspective provided by good individual or group supervision seems essential to consistent, high-quality counseling.

43 DON'T HIDE BEHIND TESTING

Because clients' expectations of counselors' expertise can influence the success of counseling (see Claiborn & Hanson, 1999; Strong, 1968), beginning counselors may be tempted to boost their credibility through testing (see Hummel, 1999).

> **COUNSELOR:** Your vocational test results indicate that you should be a salesperson. *(authoritatively)*
>
> **CLIENT:** Really?
>
> **COUNSELOR:** No doubt about it—look at these scores.

In this example, the counselor implies that the test is a foolproof method of selecting a career. Although testing data may be helpful in understanding clients and in designing counseling strategies, the counselor should remember to explain the limits of the test along with the results (American Psychological Association, 2002). This is particularly important when tests are administered by computer; clients tend to view computer-administered instruments as more scientific (Herr & Best, 1984).

Counselor credibility comes with the confidence produced by experience and knowledge, not props (Meier, 2001). Being an expert does not mean that you are dogmatic or authoritarian (Cormier & Cormier, 1998).

CLIENT: So which one of these careers on the computer list should I choose?

COUNSELOR: Well, this inventory wasn't designed to make a choice for you. This list should help you get an idea of what kinds of careers might be best for you to explore.

CLIENT: Oh.

Tests are tools. They are imperfect and useful at the same time. Test results, in and of themselves, never dictate action. Counselors employ tests as one more technique in their repertoire for helping clients.

44 ON ETHICAL QUESTIONS, CONSULT

Counselors inevitably find themselves in gray zones. Issues of counselors' duty to warn, dual relationships, confidentiality and privacy, client rights, and personal relationships with clients are subjects with considerable potential for ambiguity and risk (Kenyon, 1998; Pope & Vasquez, 1998; Welfel, 2005). In addition to knowing the formal ethical standards of your profession (for example, American Counseling Association, 2005; American Psychological Association, 2002; American Rehabilitation Counseling Association, 2002; American School Counselor Association, 2004; Association for Specialists in Group Work, 1989; Canadian Psychological Association, 2000; Herlihy & Corey, 2005; National Association of School Psychologists, 1997; National Board for Certified Counselors, 1989; National Career Development Association, 2003), you must be willing to discuss ethical issues with supervisors and colleagues.

COUNSELOR 1: He started to talk about wanting to hurt his ex-wife.

COUNSELOR 2: Did he mention a method?

COUNSELOR 1: I didn't think to ask about that. I'm wondering if I have to notify her about his threat.

Although some confusion remains, the *Tarasoff* and *Tarasoff II* cases (see Bersoff, 2003; Monahan, 1995; Schmidt & Meara, 1984; Slovenko, 1988; Stanard & Hazler, 1995) set precedents that counselors must take reasonable care (which varies case by case) to protect persons who may be physically harmed by their client. This protection could include warning the potential victim or notifying the police (Slovenko, 1988). Such actions may obviously break confidentiality, so counselors must carefully assess the seriousness of threats made by clients and explain to clients the limits to confidentiality. In addition, you also have the obligation to protect clients' lives if they are suicidal and to report child abuse if you learn of it during counseling.

New counselors also express concern about the limits of their skills. Counselors refer clients whom they cannot help. New counselors may have difficulty deciding if they have the necessary skills. Consult with your supervisor when you begin to question whether you can help a particular client.

Your supervisor may teach you new skills or help you find other assistance. Finally, an important new law has delineated regulations for client privacy. The Health Insurance Portability and Accountability Act of 1996 (HIPAA) contains requirements (only recently interpreted and implemented by the Department of Health and Human Services) that determine how personal health care information, including information about counseling, can be kept and communicated. Client confidentiality has always been a priority for counselors and has always been governed by state law. Along with these individual state laws, HIPAA sets the standard for maintaining privacy.

HIPAA provides guidelines for the physical environment of a counseling office. For example, a fax machine must be secured so that only those who have permission can see any information that may be received or sent by fax. HIPAA also requires that electronically transmitted client information be encrypted to ensure privacy. Finally, HIPAA describes how and when client information can be used and disclosed. Because the regulations are so complicated and vary from state to state, counselors should consult with their state's professional association or national websites to clarify the law in their region (for example, American Psychological Association: www.apa.org). The complete text of HIPAA regulations can be found at: www.aspe.hhs.gov.

On ethical questions, don't get caught out on a limb. Always consult if you have any doubt. Colleagues and supervisors may be more objective and able to help you decide on an appropriate course of action.

SUMMARY AND DISCUSSION QUESTIONS

As noted in Chapter 3, the counselor's assumptions and attitudes can sometimes be obstacles to helping clients. This chapter also emphasized that a counselor's knowledge of self can influence the counseling process. This includes the counselor's motivation for helping as well as comfort with affective states such as intense emotions, transference, and countertransference. Supervision can be an effective method for dealing with these issues, and beginning counselors will benefit from learning how to be open to discussions in supervision that can be emotionally challenging.

To help you understand and apply the information in this chapter, consider these questions:

Which emotions in others are you most comfortable with? Least comfortable?

Some supervisors approach supervision as a kind of counseling with their supervisees. How might this style work for you?

How might participating in counseling yourself benefit your work as a counselor?

In what graduate courses have you discussed ethical issues that relate to counseling? What were those issues?

A Brief Introduction to Intervention

CHAPTER **6**

After you build a relationship with a client, what do you do? Although many counselors emphasize the importance of relationship building in counseling, most emphasize the necessity of further intervention for change to occur (Egan, 2001). How to intervene effectively continues to separate contemporary counselors and psychotherapists. This chapter summarizes the major contemporary approaches to intervention.

Approaches to counseling and psychotherapy traditionally have focused on one of three intrapsychic domains: affect, cognition, and behavior. *Affect* refers to the feelings we experience and express (such as anger or sadness); *cognitions* are the thoughts we think (such as, "It's not my fault I failed the test"); and *behaviors* are the overt acts we do (such as smoke three packs of cigarettes a day). Clients frequently report that they seek counseling to change painful feelings such as anxiety or depression, and recent research indicates that depression and anxiety show larger changes after counseling than any other domain (Meier & Vermeersch, 2007). Counselors and psychotherapists with different orientations disagree vigorously about which of the three systems should be targeted for intervention (or at least, targeted first).

The task becomes further complicated by controversy about permanence of change (Mahoney, 1987) and the desynchrony of the systems (Hodgson & Rachman, 1974; Rachman & Hodgson, 1974). *Permanence of change* relates to the question of symptom substitution: If we change behavior without attention to underlying feelings or beliefs, will the problem resurface elsewhere? *Desynchrony* refers to the commonly observed lack of correspondence among measures of individuals' affect, cognitions, and behaviors. For example, even if clients change their behavior, weeks may pass before the corresponding feeling change occurs.

Research findings support the general efficacy of counseling, indicating that the average client who receives treatment enjoys more improvement than do two-thirds of persons who do not receive counseling (Landman & Dawes, 1982; M. Smith & Glass, 1977). Although reviews of outcome studies generally negate general claims of overwhelming superiority for any one approach (Bergin & Garfield, 1994; Wampold et al., 1997), evidence does exist for the superiority of particular therapeutic methods with some problems/clients. For example, social learning approaches have been shown to be superior to other approaches with inpatient clients (Paul & Menditto, 1992), whereas cognitive and behavioral methods appear more effective with impulsive clients who are depressed and anxious (Beutler & Harwood, 2000).

Counselors still grapple with the question of which approach or counselor works best with which client or problem (compare Krumboltz, 1966, with Glidden-Tracey & Wagner, 1995). Another general finding worth emphasizing is that the more severe the client's symptoms, the less likely that client will improve. Kendall and associates (1992) noted that counselors often fail to properly attribute severity of problems as a reason for failure to make progress.

Beginning counselors should become familiar with the basic theory and practice of many approaches. Only then can you make the informed choices necessary to create, integrate, and structure your method with any particular client. Theory helps the counselor decide how to use the elements in this book. That is, counselor actions and their timing and sequence, ideally, are at last partially made by rational choice. A theory provides a basis for making such choices and increases the likelihood that they will be of help to the client (compare to Meier, 2003).

The sections following the "Basic Counseling Texts" section contain basic counseling texts and references for well-known counseling approaches and provide a reasonable starting point for exploring the field.

45 BASIC COUNSELING TEXTS

The following texts present overviews of various counseling approaches, counseling professions, and basic counseling skills. Elaboration of the content of *The Elements of Counseling* can be found in these books.

REFERENCES

Corey, G. (2008). *Theory and practice of counseling and psychotherapy* (8th ed.). Pacific Grove, CA: Brooks/Cole.

Corsini, R., & Wedding, D. (2007). *Current psychotherapies* (8th ed.). Pacific Grove, CA: Brooks/Cole.

Egan, G. (2009). *The skilled helper: A problem-management approach to helping* (9th ed.). Pacific Grove, CA: Brooks/Cole.

Ivey, A. (2006). *Intentional interviewing and counseling* (6th ed.). Pacific Grove, CA: Brooks/Cole.

Osipow, S., Walsh, W. B., & Tosi, D. (1984). *A survey of counseling methods*. Homewood, IL: Dorsey Press.

Sampson, J. P., Reardon, R. C., Peterson, C. W., & Lenz, J. G. (2004). *Career counseling & services*. Pacific Grove, CA: Wadsworth.

Sommers-Flanagan, J., & Sommers-Flanagan, R. (2004). *Counseling & psychotherapy theories in context & practice*. New York: Wiley.

Waldinger, R. (1986). *Fundamentals of psychiatry*. Washington, DC: American Psychiatric Press.

46 PERSON-CENTERED COUNSELING

Person-centered, client-centered, or nondirective therapy centers on the work of Carl Rogers. This approach is most unique in its emphasis on clients' ability to determine relevant issues and to solve their problems. Person-centered counselors tend to see their clients positively (since all people are assumed to be striving for self-actualization) and to respond to clients with warmth, support, unconditional positive regard, genuineness, and empathy. Client-centered counselors assist clients in the change process by focusing on congruence and affect. The counselor notices client feelings and empathizes with those feelings to help clients fully experience their affect and become more open to their life experiences.

REFERENCES

Kahn, M. (2001). *Between therapist and client.* New York: Freeman.
Patterson, C. H. (1985). The therapeutic relationship: Foundations for an eclectic psychotherapy. Pacific Grove, CA: Brooks/Cole.
Raskin, N., & Rogers, C. (1995). Person-centered therapy. In R. Corsini (Ed.), *Current psychotherapies* (5th ed., pp. 155–194). Itasca, IL: Peacock.
Rogers, C. (1986). Client-centered therapy. In I. Kutash & A. Wolk (Eds.), *Psychotherapist's casebook: Theory and technique in practice* (pp. 197–208). San Francisco: Jossey-Bass.
Rogers, C. (1990). *Client-centered therapy.* Boston: Houghton Mifflin.

47 BEHAVIORAL COUNSELING

Behavioral counselors tend to be the pragmatists of the counseling profession. After all, they maintain, if it is *behavior* that we ultimately want to change (be it smoking, anxiety about school performance, or depression), then it is *behavior* that we should target in counseling. Behavioral counselors focus on inappropriate learning as the source of client problems. Thus, clients may have inappropriately learned to associate social situations with anxiety, and consequently a regimen of relaxation and assertiveness training is prescribed. Behavioral counselors pay attention to reinforcement as provided in clients' environments; if a child misbehaves at home, a behavioral counselor might teach parents about how to reward that child for more appropriate behavior.

REFERENCES

Agras, W., Kazdin, A., & Wilson, G. (1979). *Behavior therapy.* San Francisco: Freeman.
Krumboltz, J., & Thoresen, C. (1976). *Counseling methods.* New York: Holt, Rinehart & Winston.
Masters, J., Burish, T., Hollon, S., & Rimm, D. (1991). *Behavior therapy: Techniques and empirical findings* (3rd ed.). San Diego: Harcourt Brace Jovanovich.
Skinner, B. F. (2001). *Beyond freedom and dignity.* New York: Knopf.
Wilson, G., & Franks, C. (Eds.). (1982). *Contemporary behavior therapy.* New York: Guilford.
Wolpe, J. (1990). *The practice of behavior therapy* (4th ed.). New York: Pergamon Press.

48 COGNITIVE, COGNITIVE/BEHAVIORAL COUNSELING, AND SOCIAL LEARNING THEORY

Counselors with a cognitive orientation represent the latest movement in the counseling profession. In one form or another, these counselors consider inappropriate thoughts to be the cause of painful feelings and harmful behavior. Counselors such as Ellis view *irrational beliefs* (that is, beliefs without empirical evidence) as the target for interventions, whereas Beck describes how selective attention, magnifying problems, and illogical reasoning can lead to depression. Cognitive and cognitive/behavioral counseling grew from the behavioral counseling movement and share a tradition of respect for applying research to practice and doing counseling research. Cognitive therapies have been found to be among the most effective approaches for helping clients with depression, generalized anxiety, phobias, panic disorders, and obsessive-compulsive disorders (Sexton et al., 1997).

In social learning theory, on which much cognitive and cognitive/behavioral counseling is based, special emphasis is placed on individuals' learned expectations. As a result of their social experiences, individuals learn to expect that (a) some events are more personally rewarding than others; (b) certain behaviors can produce desired events, although there may be events that are uncontrollable; and (c) people differ in their feelings of competence for doing the behaviors that can produce desired events. Although social learning theorists view expectations as the driving force, these counselors often do not directly intercede with cognitions. Instead, they modify clients' expectations through actual performance (such as in a gradual exposure to increasingly fearful situations, as is done with snake phobics) or through the use of models who demonstrate skilled behavior.

REFERENCES

Bandura, A. (1969). *Principles of behavior modification*. New York: Holt, Rinehart & Winston.

Bandura, A. (1977). Self-efficacy theory: Toward a unifying view of behavioral change. *Psychological Review, 84,* 191–215.

Bandura, A. (1997). *Self-efficacy: The exercise of control*. New York: Freeman.

Beck, A. (1979). *Cognitive therapies and the emotional disorders*. New York: International Universities Press.

Cormier, W., & Cormier, L. (1997). Interviewing strategies for helpers: Fundamental skills and cognitive behavioral interventions (3rd ed.). Pacific Grove, CA: Brooks/Cole.

Ellis, A. (1998). Rational emotive behavior therapy: A therapist's guide. San Luis Obispo, CA: Impact.

Ellis, A., & Harper, R. (1976). *A new guide to rational living*. North Hollywood, CA: Wilshire.

Emery, G., Hollon, S., & Bedrosian, R. (Eds.). (1981). *New directions in cognitive therapy*. New York: Guilford.

Hollon, S. D., & Beck, A. T. (1994). Cognitive and cognitive-behavioral therapies. In A. E. Bergin & S. L. Garfield (Eds.), *Handbook of psychotherapy and behavior change* (4th ed., pp. 428–467). New York: Wiley.

Kelly, G. (1992). The psychology of personal constructs. London: Routledge.

Mahoney, M. J., & Lyddon, W. J. (1988). Recent developments in cognitive approaches to counseling and psychotherapy. *The Counseling Psychologist, 16,* 190–234.

Meichenbaum, D. (1977). Cognitive behavior modification: An integrative approach. New York: Plenum.

Meichenbaum, D. (1999). The evolution of a cognitive-behavior therapist. In J. K. Zeig (Ed.), *The evolution of psychotherapy: The third conference* (pp. 95–104). New York: Brunner/Mazel.

Mischel, W., Shoda, Y., & Ayduk, O. (2007). *Introduction to personality* (8th ed.). New York: Holt, Rinehart & Winston.

Rosenthal, T., & Zinimennan, B. (1978). *Social learning and cognition*. New York: Academic Press.

Rotter, J. (1971). *Clinical psychology* (2nd ed.). Englewood Cliffs, NJ: Prentice-Hall.

Rotter, J. (1982). The development and applications of social learning theory: Selected papers. New York: Praeger.

Rotter, J. (1987). *Social learning and clinical psychology*. Englewood Cliffs, NJ: Prentice-Hall.

Rotter, J., Chance, J., & Phares, E. (1972). *Applications of a social learning theory of personality*. New York: Holt, Rinehart & Winston.

Williamson, E. (1959). Some issues underlying counseling theory and practice. In W. Dugan (Ed.), *Counseling points of view*. Minneapolis: University of Minnesota Press.

49 GESTALT COUNSELING

Fritz Perls is the counselor most strongly associated with Gestalt counseling. Perls frustrated clients to help them move toward self-support and away from therapist support. Gestalt counselors also emphasize body movement as a method of experiencing feelings and facilitating psychological growth. As do person-centered counselors, Gestalt counselors pay particular attention to noticing client feelings, staying in the here-and-now, and avoiding intellectual analysis of problems.

References

Downing, J. (1976). *Gestalt awareness.* New York: Harper & Row.

Goodman, P., Perls, F., & Hefferline, R. (1994). *Gestalt therapy: Excitement and growth in the human personality.* Highland, NY: Gestalt Journal Press.

Hardy, R. E. (1991). *Gestalt psychotherapy.* Springfield, IL: Thomas.

Perls, F. (1970). *In and out of the garbage can.* New York: Bantam.

Perls, F. (1992). *Gestalt therapy verbatim.* Highland, NY: Gestalt Journal Press.

Perls, F., Hefferline, R., & Goodman, P. (1977). *Gestalt therapy: Excitement and growth in the human personality.* New York: Bantam.

Polster, E., & Polster, M. (1974). *Gestalt therapy integrated.* New York: Brunner/Mazel.

Zinker, J. (1978). *Creative process in Gestalt therapy.* New York: Random House.

50 PSYCHOANALYTIC AND PSYCHODYNAMIC COUNSELING

Sigmund Freud established the foundation from which all counseling approaches evolved. His ideas about the unconscious and about personality development led to ingenious counseling techniques and motivated opponents to create such radically different approaches as behavioral counseling and rational/emotive therapy. Much of the work of the psychoanalytic counselor involves making unconscious material conscious, thereby helping counselor and client to gain insight into the mechanisms of psychological adjustment.

This process, however, is frequently anxiety provoking to clients who resist self-awareness by means of various defense mechanisms. More than do adherents of any other approach, psychoanalytic counselors attempt to understand the client through the client-therapist relationship as it resembles earlier transactions between the client and authority figures (for example, parents). The projection onto the counselor of clients' feelings toward their parents is called *transference.* By bringing transference into the open, clients gain new understanding of their psychological processes and ameliorate troubling symptoms.

References

Basch, M. (1992). *Practicing psychotherapy.* New York: Basic Books.

Brill, A. (Ed.). (1985). *The basic writings of Sigmund Freud.* New York: Modern Library.

Chapman, A. H. (1995). The treatment techniques of Harry Stack Sullivan. Northvale, NJ: Aronson.

Clark, A. J. (1998). Defense mechanisms in the counseling process. Thousand Oaks, CA: Sage.

Freud, A. (1990). *The ego and the mechanisms of defense.* New York: International Universities Press.

Jung, C. (1990). *Basic writings.* New York: Modern Library.

Kernberg, O. (1993). Object relations theory and clinical psychoanalysis. Northvale, NJ: Aronson.

Langs, R. (1973). *The techniques of psychoanalytic psychotherapy* (Vol. 1). Northvale, NJ: Aronson.

Malcolm, J. (1994). *Psychoanalysis: The impossible profession.* Northvale, NJ: Aronson.

Menninger, K., & Holzman, P. (1995). *Theory of psychoanalytic technique* (2nd ed.). Northvale, NJ: Aronson.
St. Clair, M. (1987). Object relations and self-psychology: An introduction. Pacific Grove, CA: Brooks/Cole.
Shapiro, D. (1999). *Neurotic styles*. New York: Basic Books.
Strupp, H., & Binder, J. (1990). Psychotherapy in a new key: A guide to time-limited dynamic psychotherapy. New York: Basic Books.
Sullivan, H. S. (1953). The collected works of Harry Stack Sullivan, M.D. New York: Norton.
Wachtel, P. (1990). Psychoanalysis and behavior therapy: Toward an integration. New York: Basic Books.

51 EXISTENTIAL COUNSELING

Existential counselors examine the role of what many consider to be abstract, philosophical issues in the psychological lives of individuals. More than counselors using any other approach, existential counselors eschew technique in favor of grappling with the basic dimensions of life and death.

Thus, people can be considered in terms of *being* (which corresponds to an awareness of oneself) and *nonbeing* (a loss of identity, perhaps caused by conformity). Rollo May, an eminent existential counselor, describes anxiety as the experience of the threat of imminent nonbeing. As do person-centered and Gestalt counselors, existentialists see personal choice and volition as basic facts of human existence.

According to existential counselors, clients seek counseling to expand their psychological worlds. As an existential counselor, one's job is to be authentic, to expose oneself to clients so that the clients can become aware of similar qualities in themselves. Along with many behaviorists, existential counselors believe that knowledge and insight follow behavior change, not vice versa.

REFERENCES

Frankl, V. (2000). *Man's search for meaning* (3rd ed.). New York: Washington Square Press.
Jourard, S. (1971). *The transparent self* (Rev. ed.). New York: Van Nostrand Reinhold.
Keen, E. (1970). Three faces of being: Toward an existential clinical psychology. New York: Appleton-Century-Crofts.
May, R. (1995). *The discovery of being*. New York: Norton.
Sartre, J. P. (1993). *Being and nothingness*. New York: Washington Square Press.
Yalom, I. (2000). Love's executioner and other tales of psychotherapy. New York: Basic Books.

52 GROUP COUNSELING

Group counselors may subscribe to any of the counseling approaches described previously; that is, you may find Gestalt groups, person-centered groups, and cognitive-behavioral groups. What all group counselors share in common is a recognition of the benefits of working with more than one client at a time. If clients can be placed in a group on the basis of relatively similar problems, work can be done more efficiently than it can in one-to-one counseling. In most groups, benefits occur from interactions among group members, not from working with the counselor per se. With therapeutic groups, the group counselor is more a facilitator than an active participant or leader. Skilled group counselors supply moderate amounts of group rules and emotional challenges, along with high amounts of support and interpretation of group processes (Lieberman, Yalom, & Miles, 1973).

References

Bednar, R. L., & Kaul, T. J. (1994). Experiential group research: Can the canon fire? In A. E. Bergin & S. L. Garfield (Eds.), *Handbook of psychotherapy and behavior change* (4th ed., pp. 631–663). New York: Wiley.

Corey, G. (2007). *Theory and practice of group counseling* (7th ed.). Pacific Grove, CA: Brooks/ Cole.

DeLucia-Waack, J. L., Gerrity, D. A., Kalodner, C. R., & Riva, M. (2004). *Handbook of group counseling and psychotherapy*. Thousand Oaks, CA: Sage.

Fuhriman, A., & Burlingame, G. M. (1994). *Handbook of group psychotherapy*. New York: Wiley.

Gazda, G. (1990). *Group counseling: A developmental approach* (4th ed.). Boston: Allyn & Bacon.

Gladding, S. (2002). *Group work: A counseling specialty* (3rd ed.). New York: Merrill/Prentice-Hall.

Schneider Corey, M., & Corey, G. (2006). *Groups: Process and practice* (7th ed.). Pacific Grove, CA: Brooks/Cole.

Yalom, L. (2005). *The theory and practice of group psychotherapy* (5th ed.). New York: Basic Books.

53 FAMILY/SYSTEMS COUNSELING

Family and systems approaches are relatively new approaches to counseling. In contrast to traditional methods, which focus on processes within the individual as the target of change, family and systems counselors chart the influence of social systems on clients. The goal of the family counselor is to detect and understand the methods that family members use to communicate among themselves, maintain the family structure, and help or hinder members' growth. Many systems counselors share with Gestalt counselors an interest in psycholinguistics, the study of how language influences what we think, do, and feel.

References

Bandler, R., & Grinder, J. (1982). *The structure of magic*. Palo Alto, CA: Science and Behavior Books.

Bateson, G. (1999). *Steps to an ecology of mind*. Northvale, NJ: Aronson.

Becvar, D., & Becvar, R. (2008). *Family therapy* (7th ed.). Boston: Allyn & Bacon.

Bowen, M. (1994). *Family therapy in clinical practice*. Northvale, NJ: Aronson.

deShazer, S. (1982). *Patterns of brief family therapy*. New York: Guilford.

Haley, J. (1991). *Problem-solving therapy* (2nd ed.). San Francisco: Jossey-Bass.

Madanes, C. (1991). Behind the one-way mirror: Advances in the practice of strategic therapy. San Francisco: Jossey-Bass.

Mikesell, R. H., Lusterman, D., & McDaniel, S. H. (Eds.). (1995). *Integrating family therapy*. Washington, DC: American Psychological Association.

Minuchin, S., & Fishman, H. C. (2004). *Family therapy techniques*. Cambridge, MA: Harvard University Press.

Minuchin, S., & Nichols, M. P. (2001). Structural family therapy. In F. M. Dattilio (Ed.), *Case studies in couple and family therapy: Systematic and cognitive perspectives* (pp. 108–131). New York: Guilford.

Napier, A., & Whitaker, C. (1988). *The family crucible*. New York: HarperCollins.

Nichols, M., & Schwartz, R. (2007). *Family therapy: Concepts and methods* (8th ed.). Boston: Allyn & Bacon.

Satir, V., & Baldwin, M. (1984). *Satir step by step*. Palo Alto, CA: Science and Behavior Books.

Watzlawick, P., Weakland, J., & Fisch, R. (1974). *Change: Principles of problem formation and problem resolution*. New York: Norton.

Worden, M. (2002). *Family therapy basics* (3rd ed.). Pacific Grove, CA: Brooks/Cole.

54   MULTICULTURAL COUNSELING

Multicultural counseling refers to approaches that emphasize cultural, ethnic, and racial influences on counseling process and outcome (Parham, 2002). Proponents of multicultural counseling describe a range of philosophies and methods but suggest that at a minimum, counselors should (a) be aware of their values and potential biases with clients of color and (b) attend to potential influences on counseling process and outcome brought by the client's and counselor's cultural background. Regarding process, for example, some evidence suggests that counselors' biases about clients can be passed on when they interpret tests (Fouad & Chan, 1999).

Culturally specific knowledge may be most helpful to counselors when they work regularly with a few different cultural groups. For example, counselors at a college counseling center who have a significant caseload that includes Asian international students would be well advised to know that research suggests that such clients' perceptions of empathy increase when counseling focuses on social and family causes of problems rather than intrapsychic influences (D. Sue & Sundberg, 1996). In general, counselors should obtain and respond to culturally specific knowledge with their clients (Fischer et al., 1998). Fischer and associates suggest that with clients of color, counselors apply what they know about counseling in a culturally sensitive manner. They recommend a balance between attention to a client's cultural background and to the uniqueness of each client. Fischer et al. note that race and culture will vary in importance for clients of color (compare to Ridley, 2005). For some highly acculturated clients, culture may hold little significance in their worldview; for other clients, the counselor's ability to appreciate, know, and apply cultural meanings will be the key to effective helping.

REFERENCES

Abreu, J. M. (2001). Theory and research on stereotypes and perceptual bias: A didactic resource for multicultural counseling trainers. *The Counseling Psychologist, 29,* 487–512.

Fischer, A. R., Jome, L. M., & Atkinson, D. R. (1998). Reconceptualizing multicultural counseling: Universal healing conditions in a culturally specific context. *The Counseling Psychologist, 26,* 525–588.

Ridley, C. R. (2005). Overcoming unintentional racism in counseling and therapy (2nd ed). Thousand Oaks, CA: Sage.

Parham, T. A. (Ed.). (2002). *Counseling persons of African descent.* Thousand Oaks, CA: Sage.

Patterson, C. H. (1996). Multicultural counseling: From diversity to universality. *Journal of Counseling & Development, 74,* 227–231.

Pedersen, P. B., Draguns, J. G., Lonner, W. J., & Trimble, J. E. (Eds.). (2002). *Counseling across cultures* (5th ed.). Thousand Oaks, CA: Sage.

Ponterotto, J. G., Casas, J. M., Suzuki, L. A., & Alexander, C. M. (Eds.). (2001). *Handbook of multicultural counseling.* Thousand Oaks, CA: Sage.

Sue, D. W., & Sue, D. (2007). *Counseling the culturally diverse: Theory and practice* (5th ed.). New York: Wiley.

Sue, D. W., Ivey, A. E., & Pedersen, P. B. (Eds.). (1996). *A theory of multicultural counseling and therapy.* Pacific Grove, CA: Brooks/Cole.

Whaley, A. L. (2001). Cultural mistrust and mental health services for African Americans: A review and meta-analysis. *The Counseling Psychologist, 29,* 513–531.

55   FEMINIST THERAPY

Feminist counselors seek an awareness of how their personal values influence counseling and, with many clients, openly share their feminist orientation. Although feminist authors have offered a range of philosophical beliefs and therapeutic approaches, a common core of five values and attitudes can be described.

The first emphasis focuses on how sex-role socialization influences personal and vocational choices. For example, a woman labeled as dependent may have learned as a child that being a good woman meant that she should be passive. Similarly, a woman who acts assertively may run the risk of being labeled angry or selfish. Feminist therapists seek to place these and other characteristics and behaviors in the social and political context of societies where maleness is viewed as the healthy norm. Similarly, a feminist counselor would be more likely to encourage a female high school student interested in health care to consider becoming a physician (instead of the traditional sex-role job of nurse). In therapy, clients may perform a sex-role analysis in which they consider how social expectations about gender influence their lives (Worell & Remer, 2002).

Feminist counselors also tend to locate causes of psychological problems in environmental rather than intrapsychic conditions. Feminist therapists avoid blaming the victim. A depression experienced by a woman in an abusive relationship would be seen as a result of that relationship, rather than evidence of a psychological weakness in the client. This client would likely be encouraged to change or leave the abusive relationship instead of focusing simply on an intrapsychic adjustment to alleviate the depression.

A third feminist value involves engaging in social change actions that might improve women's psychological and physical health. If the important causes of women's difficulties are rooted in the political and social environment, it makes sense to focus change efforts there. Surveys indicate that more women than men present with such problems as depression, anxiety, eating disorders, and agoraphobia; in turn, women evidence greater utilization of mental health services (Worell & Remer, 2002). The phrase "The personal is political" refers to the belief that women's problems are rooted in their social and political culture (Enns, 1997). For example, the fact that women in many occupations are paid less than men can contribute to women's stress, financial and otherwise; thus, a feminist therapist would likely support (and encourage clients to support) political action toward equal pay. Some feminist therapists, in fact, see therapy "as an act of political resistance" (Brown, 2000, p. 378).

Fourthly, feminist counselors emphasize equality in human relationships. Feminist therapists are particularly aware of power differences in human relationships and systems. In an attempt to move toward equality in the counselor-client relationship, feminist therapists demystify the therapy experience, for example, by sharing their assessments about clients in clear, jargon-free language (Enns, 1997; Worell & Remer, 2002). Feminist therapists also avoid making decisions for clients and communicate confidence in the client's competence (Rawlings & Carter, 1977).

And finally, feminist counselors work to raise personal and social consciousness about women's issues. Feminist therapists believe that recognition of these values and beliefs can be important outcomes in counseling (Enns, 1997). Similarly, feminist therapists may attend and encourage their clients to attend consciousness-raising groups where women share their experiences and discuss how social/cultural factors influence them.

Whereas feminist counselors draw upon other therapeutic approaches for techniques (particularly Rogerian), feminist values and beliefs affect how these methods are employed with clients. The central role of the client's gender and related characteristics (such as race and culture) are thus explored in a way that enhances counseling progress and the client's mental health.

REFERENCES

Brabeck, M. M. (Ed.). (2000). *Practicing feminist ethics in psychology*. Washington, DC: American Psychological Association.

Broverman, L., Broverman, D., Clarkson, E., Rosenkrantz, P., & Vogel, S. (1970). Sex-role stereotypes and clinical judgments of mental health. *Journal of Consulting and Clinical Psychology, 34*, 1–7.

Brown, L. (2004). Subversive dialogues: Theory in feminist therapy. New York: Basic Books.

Brown, L. (in press). Feminist therapy. Washington, DC: American Psychological Association.

Fouad, N. A., & Chan, P. M. (1999). Gender and ethnicity: Influence on test interpretation and reception. In J. W. Lichtenberg & R. K. Goodyear (Eds.), *Scientist-practitioner perspectives on test interpretation* (pp. 31–58). Boston: Allyn & Bacon.

Garner, J. D., & Enns, C., & (2004). *Feminist theories and feminist psychotherapies*. New York: Haworth.

Rawlings, E. I., & Carter, D. K. (1977). Feminist and nonsexist psychotherapy. In E. I. Rawlings & D. K. Carter (Eds.), *Psychotherapy for women* (pp. 49–76). Springfield, IL: Thomas.

Worell, J., & Remer, P. (2002). *Feminist perspectives in therapy* (2nd ed.). New York: Wiley.

56 BRIEF THERAPY AND SOLUTION FOCUSED THERAPY

Brief therapy (BT) can be considered one of the newest and most ambiguous approaches to counseling. BT's increased use has been significantly influenced by economics—that is, because of the widespread implementation of managed care approaches whose major emphasis is limiting costs. However, BT is ambiguous along more than one dimension: (a) Proponents of different approaches argue that BT can be comprised of one session or weekly sessions over several years, and (b) BT often appears to be simply a shortened version of other counseling approaches. What does seem common across its different forms is that BT counselors plan sessions more than other counselors and are more active and directive.

Reduction of symptoms (for example, psychological distress) and restoration of functioning are primary goals; this therapy is most appropriate for highly motivated clients with clearly defined complaints. Although BT by definition focuses on time limits and efficiency, there is also a recognition that clients return periodically to counseling. Budman (1983) suggests that problems such as depression, stress, and eating disorders are chronic, persistent, and intermittent in nature. E. Frank (1991) reports that depressed patients, treated in 16 sessions and continued intermittent contact, had fewer relapses than those with only the 16 initial sessions. Thus, Budman maintains that brief therapies emphasize efficiency, outcomes, and benefits of provided services.

Counselors have increasingly turned their attention to solution-focused therapy (SFT), a type of BT that turns traditional assumptions on their head in a way that many counselors find useful. In contrast to problem-focused therapies, SFT emphasizes client strengths and focuses on the client's current and future goals; clients are assumed to have all the resources they need to change (Walter & Peller, 1992). Thus, the SFT therapist will ask "What would you like to change?" rather than inquire in depth about past problems and history (O'Connell, 2005). Counselor and client begin with the client's presenting trouble(s) but may negotiate to prioritize a key issue to be addressed in therapy. SFT counselors typically seek to do the minimal and simplest intervention possible in order to minimize therapy length and client dependency (O'Connell, 2005). A small change can create hope and lead into bigger changes, and examples of times when the problem was absent could be examined and amplified (Quick, 1996). SFT counselors also encourage clients to do more of what's working for them and less of what is not working (Quick, 1996). Work between sessions by the client can be equally as important or more important than what transpires within a session.

The most obvious question about types of BT is this: Do they work? Research examining the effects of the number of counseling sessions on outcome has provided no clear consensus. A *Consumer Reports* survey ("Does Therapy Help?," November 1995) of 4,000 readers who had previous counseling found that a greater number of sessions were associated with more improvement. But other reviews have suggested no differences or advantages for BT (see Barber, 1994; Steenbarger, 1994). As with most areas in counseling, BT requires more study.

Research suggests that different sorts of problems might be amenable to short- and long-term counseling. In a study of SCL-90-R items (a symptom checklist) completed by psychotherapy outpatients, Kopta, Howard, Lowry, and Beutler (1994) found that the items could be grouped differentially on the basis of the amount of change displayed in response to treatment. Kopta and et al.'s analyses classified the items into three categories: (a) acute (quick response to treatment), (b) chronic distress (moderate response rate), and (c) characterological (slow response rate). Thus, clients reported relatively quick improvement on items assessing acute distress symptoms (such as temper outbursts and hopelessness), whereas characterological items (such as paranoia and sleep trouble) took longer to evidence change. Kopta and others concluded that a typical client required about one year of outpatient treatment to have a 75 percent chance of symptomatic recovery.

REFERENCES

Atlas, J. A. (1994). Crisis and acute brief therapy with adolescents. *Psychiatric Quarterly, 65*, 79–87.

Barber, J. P. (1994). Efficacy of short-term dynamic psychotherapy: Past, present, and future. *Journal of Psychotherapy Practice and Research, 3*, 108–121.

Budman, S. H. (1990). Brief therapy in the year 2000 and beyond: Looking back while looking forward. In S. H. Budman (Ed.), *Forms of brief therapy* (pp. 461–470). New York: Guilford.

Crits-Christoph, P. (1992). The efficacy of brief dynamic psychotherapy: A meta-analysis. *American Journal of Psychiatry, 149*, 151–158.

Cummings, N., & Sayama, M. (1995). Focused psychotherapy: A casebook of brief, intermittent psychotherapy throughout the life cycle. New York: Brunner/Mazel.

Davanloo, H. (Ed.). (1995). *Short-term dynamic psychotherapy*. Northvale, NJ: Aronson.

Does therapy help? (1995, November). *Consumer Reports, 60*, 734–739.

Frank, E. (1991). Interpersonal psychotherapy as a maintenance treatment for patients with recurrent depression. *Psychotherapy, 28*, 259–266.

Koss, M. P., & Shiang, J. (1994). Research on brief psychotherapy. In A. E. Bergin & S. L. Garfield (Eds.), *Handbook of psychotherapy and behavior change* (4th ed., pp. 664–700). New York: Wiley.

Kreider, J. W. (1998). Solution-focused ideas for briefer therapy with longer-term clients. In M. F. Hoyt (Ed.), *Handbook of constructive therapies: Innovative approaches from leading practitioners* (pp. 341–357). San Francisco: Jossey-Bass.

Malan, D. H. (1976). *The frontier of brief psychotherapy*. New York: Plenum.

McCullough, L., Winston, A., Farber, B., Porter, F., Pollack, J., Laikin, M., et al. (1991). The relationship of patient-therapist interaction to outcome in brief psychotherapy. *Psychotherapy, 28*, 525–533.

Miller, G., & deShazer, S. (1998). Have you heard the latest rumor about . . . ? Solution-focused therapy as rumor. *Family Process, 37*, 363–377.

Murphy, J. J. (1994). Brief therapy for school problems. *School Psychology International, 15*, 115–131.

O'Connell, B. (2005). *Solution-focused therapy* (2nd ed.). London: Sage.

Quick, E. K. (2008). Doing what works in brief therapy: A strategic solution focused approach (2nd ed.). San Diego: Academic Press.

Steenbarger, B. N. (1994). Duration and outcome in psychotherapy: An integrative review. *Professional Psychology Research and Practice, 25,* 111–119.

Strupp, H., & Binder, J. (1990). Psychotherapy in a new key: A guide to time-limited dynamic psychotherapy. New York: Basic Books.

Talmon, M. (1990). *Single session therapy.* San Francisco: Jossey-Bass.

Walter, J. L., & Peller, J. E. (1992). *Becoming solution-focused on brief therapy.* New York: Brunner/Mazel.

57 INTEGRATIVE APPROACHES

Rather than fight battles over which counseling approach is best, some counselors have become interested in discerning common elements in different therapeutic approaches and theories (Ivey, 2002; Staats, 1983; Zeig, 1997). Counseling theorists have proposed a number of integrated approaches (Stricker & Gold, 2001), but most can be placed along a continuum of *technical eclectism* (where the counselor makes intuitive guesses about therapeutic direction, without theory to guide decisions) to *theoretical integration* (where theory fully guides the choice of therapeutic method) (Petrocelli, 2002). We will briefly describe two integrative approaches that have a strong theoretical basis.

Proponents of a *common factors approach* point to the findings of psychotherapy outcome studies that indicate that all approaches have approximately equal effects (for example, M. Smith & Glass, 1977; Wampold et al., 1997). More recent studies have also found that clients may evidence improvement in counseling even before the active intervention is introduced (compare Kelly, Roberts, & Ciesla, 2005; Lambert, 2005). Frank and Frank (1991) suggest that the benefits of counseling derive not simply from specific techniques associated with different schools, but from elements common to all therapeutic approaches. These common elements are:

1. Developing a therapeutic relationship, that is, an emotionally engaged alliance between counselor and client.
2. Creating or discovering a shared worldview, for example, about the causes and context of the client's problems.
3. Increasing the client's expectations for improvement (that is, hope).
4. Choosing or creating an intervention that both client and counselor believe will be effective.

Many of the concepts presented in *The Elements of Counseling* can be related to these four components. For example, the principle of listening closely to what clients say (Chapter 2) constitutes a procedure one can use in learning the client's worldview.

Another integrative approach focuses on proposed *common stages of change* in therapy. These are sequences of events that, in theory, most clients proceed through on their way to successful outcomes. For example, Prochaska and DiClemente's (1983) transtheoretical approach examines three fundamental dimensions of change: (a) processes, (b) stages, and (c) levels. Prochaska (1995; see also Prochaska & Norcross, 2006) describes psychotherapeutic change as unfolding over six stages:

1. Precontemplation, when a client has little or no awareness about the major problem.
2. Contemplation, when a client becomes aware of the problem and begins to think about how to remedy it.

3. Preparation, when a client intends to change and begins to take preliminary actions.
4. Action, when a client changes behavior or the environment in an attempt to remedy the problem.
5. Maintenance, when a client attempts to continue the changes made.
6. Termination, when the problem is fully resolved and the client is confident it will not reoccur.

Of particular importance in this stage theory is that different interventions can be more effective in certain stages (Prochaska, 1995). Informational workshops about eating disorders, for example, could help young women in the Precontemplation stage become more aware of their eating and dieting problems.

References

DiClemente, C. C., & Prochaska, J. O. (1985). Processes and stages of change: Coping and competence in smoking behavior change. In S. Shiffman & T. A. Wills (Eds.), *Coping and substance abuse* (pp. 319–342). New York: Academic Press.

Norcross, J. C., & Goldfried, M. R. (Eds.). (2005). *Handbook of psychotherapy integration* (2nd ed.). New York: Basic Books.

Prochaska, J. O. (1995). An eclectic and integrative approach: Transtheoretical therapy. In A. S. Gurman & S. B. Messer (Eds.), *Essential psychotherapies* (pp. 403–440). New York: Oxford University Press.

Prochaska, J. O., & DiClemente, C. C. (1983). Stages and processes of self-change in smoking: Toward an integrative model of change. *Journal of Consulting and Clinical Psychology, 5*, 390–395.

Prochaska, J. O., & Norcross, J. (2006). *Systems of psychotherapy: A transtheoretical analysis* (6th ed.). Pacific Grove, CA: Wadsworth.

Stricker, G., & Gold, J. R. (1993). *Comprehensive handbook of psychotherapy integration*. New York: Plenum.

Velicer, W. F., Rossi, J. S., Prochaska, J. O., & DiClemente, C. C. (1996). A criterion measurement model for health behavior change. *Addictive Behaviors, 21*, 555–584.

58 NARRATIVE THERAPY

Narrative therapy (NT) refers to a variety of approaches that focus on the role of language and stories in counseling. NT invites us to notice that clients tell stories in therapy and that these stories can be useful in assessing and helping clients. According to one of the founders of NT, White, stories are powerful because all people actively seek to make sense of their lives through plausible narratives (White & Epston, 1990).

The process of narrative therapy can be difficult to describe because it can vary considerably by client and counselor. The narrative counseling process often appears relatively unique to the particular client and counselor working together, with the counselor finding methods to overcome obstacles and facilitate client progress. For example, Greenberg and Angus (2004) describe the counseling process as helping clients stay with and articulate their feelings; by doing so, key themes can fully emerge and previous habitual interpretations are constrained. Hardtke and Angus (2004), however, observe that many clients tend to stay in storytelling while therapists attempt to focus them on internal states. As a result, the counseling process typically involves weaving between (a) talking about emotional personal narratives and (b) efforts to make sense of those narratives (that is, meaning making).

Anderson (2004) notes that more emotion is typically present in the beginning of storytelling, with meaning making occurring later in the process.

Empirical support for at least one type of narrative approach can be found in dozens of research studies that have found that writing about emotional upheavals can positively affect psychological and physical health (Pennebaker et al., 2003). In these studies, individuals typically write about an emotionally disturbing topic for 15 to 30 minutes per day over a three- to five-day period. The results of writing studies often parallel counseling research findings or have potential implications for how narrative therapy should be conducted. Writing about emotional events, for example, can lead to long-term improvements but also produce short-term increases in emotional distress (Smyth, 1998). Healthy writing is associated with a high number of self-references on some days but not on others, suggesting that "people who always write in a particular voice—such as first person singular—simply do not improve" (Pennebaker et al., 2003, p. 569). This finding suggests that successful clients learn how to shift back and forth between their personal perspective and an empathic view of others' outlook.

Some research suggests that clients' sharing of narratives can strengthen the working alliance and help clients reexperience emotional events (Hardtke & Angus, 2004). Creating meaning appears to help in the regulation of emotion as well as in the maintenance and repair of interpersonal relationships (Singer & Blagov, 2004). Other studies indicate that writing about trauma can reduce the number of physician visits, improve physical health, improve academic grades, and increase reemployment rates among those who have lost jobs (Pennebaker et al., 2003). One explanation for the benefits of writing tasks and narrative construction is that "individuals who write about traumas naturally come to a coherent understanding of the event" (Pennebaker et al., 2003, p. 567). In other words, individuals produce a meaningful narrative around their problem(s), as typically occurs in many instances of successful counseling.

One of the advantages of NT is that since all therapists pay attention to language to some degree, it can be combined with almost any other counseling approach. Cognitive behavioral therapists such as Ellis, described earlier in this chapter, listen for clients' use of words such as *must* and *should* that can be indicative of irrational beliefs (for example, "I must get an A on this exam!"). Research conducted by cognitive behavioral clinicians has found that the memories of trauma victims tend to be disorganized and fragmented (Amir, Stafford, Freshman, & Foa, 1998) and that treatment with prolonged exposure therapy was associated with increased organization in the narrative around trauma (Foa, Molnar, & Cashman, 1995). Finally, some NT therapists attempt to explore the affect in client stories as a way to produce a corrective emotional experience, while the benefits of meaning making described above also fits well with existential counseling.

REFERENCES

Angus, L. E., & McLeod, J. (Eds.) (2004). *The handbook of narrative and psychotherapy.* Thousand Oaks, CA: Sage.

Bandler, R., & Grinder, J. (1975). *The structure of magic.* Palo Alto, CA: Science and Behavior Books.

Pennebaker, J. W., Mehl, M. R., & Niederhoffer, K. G. (2003). Psychological aspects of natural language use: Our words, our selves. *Annual Review of Psychology, 54,* 547–577.

Pennebaker, J. W., & Seagal, J. D. (1999). Forming a story: The health benefits of narrative. *Journal of Clinical Psychology, 55,* 1243–1254.

Polkinghorne, D. E. (2004). Narrative therapy and postmodernism. In L. E. Angus & J. McLeod (Eds.), *The handbook of narrative and psychotherapy* (pp. 53–67). Thousand Oaks, CA: Sage.
White, M., & Epston, D. (1990). *Narrative means to therapeutic ends*. New York: Norton.

59 NEW AND EMERGING APPROACHES

New and emerging approaches are counseling and psychotherapy interventions that propose innovative methods of thinking about and working with clients. New counseling approaches are continually being introduced; many disappear after a few years, but for a few others, professionals maintain or increase their interest. In this section we describe several of the newer approaches that we think will have staying power. Empirical research provides support for the general efficacy of these approaches, but like all counseling and psychotherapy methods, important questions remain.

Acceptance and commitment therapy (ACT; Hayes, Strosahl, & Wilson, 1999; Luoma, Hayes, & Walser, 2007) is an interesting combination of Eastern philosophy and the empiricism of cognitive and behavioral therapies. One of the basic assumptions is that "ordinary human psychological processes can themselves lead to extremely destructive and dysfunctional results" (Hayes et al., 1999, p. 6). Instead of the normal need for control that can sometimes lead to psychological problems, for example, clients learn mindfulness, which involves focusing on the present moment. Instead of obsessively thinking (that is, using internal analytic language) about problems, clients learn to notice and accept their thoughts. For their part, therapists focus less on the content of client's problems than on understanding and changing the context around that content. A client who anxiously obsesses about an upcoming 70-hour workweek of special meetings and presentations might be asked to explore the context of that worry. That initial exploration might reveal that for this particular individual and situation, (a) the client actually has sufficient time to prepare and do the work, indicating that the anticipatory worry is more of an issue than the actual work, (b) the client sees the anticipatory worry about the special week as an indication she does not have the necessary stamina to hold a professional position, and (c) some of the work is actually unnecessary and might be changed with negotiations with a boss or colleagues. The therapist would help the client learn to perceive these thoughts as thoughts, accept the thoughts without struggling with them, set goals in relation to what the client values in this context, and then act to fulfill the client's goals. More information about ACT may be accessed at www.contextualpsychology.org.

Young's schema therapy also combines elements shared with other approaches, including behavior therapy, cognitive therapy, object relations, and Gestalt therapy (Young, Klosko, & Weishaar, 2003). Schemas are seen as self-defeating themes that individuals continue to repeat throughout their lives. Typically these themes relate to oneself or to relationship with others and were developed during childhood. Examples of such schemas include abandonment or instability with significant others, defectiveness and shame, social isolation and alienation, and subjugation. People cope with these schemas in a variety of ways, including surrendering, avoiding, or overcompensating (that is, doing the opposite). Therapists help clients learn to recognize their predominant schemas and coping styles, often by probing into early childhood experiences, focusing on the therapeutic relationship, and other methods from the approaches listed above. More information about schema therapy can be found at www.schematherapy.com and www.cognitivetherapy.me.uk/schema_theory.htm.

One of the most interesting recent developments in counseling theory is an increasing focus on client affect as a key component of successful interventions. Basic researchers have been conducting important studies on emotion (compare Barrett, 2006; Fridja & Sundararajan, 2007; Izard, 2007), and emotion may become the central construct that links different counseling and psychotherapy approaches. For example, even more behavior-oriented researchers have recently considered how (a) individuals' maladaptive processes can increase anxiety and depression and that (b) suppression (that is, hiding feelings from self and/or others) and avoidance are examples of such processes (compare with Campbell-Sills & Barlow, 2007). One newer approach that explicitly focuses on client affect is emotion-focused therapy (Greenberg, 2008). Greenberg (2008) stated the central idea: "Whereas thinking usually changes thoughts, feeling usually changes emotion" (p. 53). Pennebaker and Beall (1986), for example, found that persons who wrote about the factual aspects of a traumatic episode did not show improvement on health variables, whereas those who write about emotional aspects did. Finally, if changes in emotions such as depression and anxiety (that is, negative affect [NA]) are concomitants of all problem resolution in counseling (Meier & Vermeersch, 2007), then a measure of NA might function well as a universal indicator of progress and outcome.

REFERENCES

Fridja, N. H., & Sundararajan, L. (2007). Emotion refinement. *Perspectives on Psychological Science*, 2, 227–241.
Giesen-Bloo, J., Van Dyck, R., Spinhoven, P., Van Tilburg, W., Dirksen, C., Van Asselt, T., et al. (2006). Outpatient psychotherapy for Borderline Personality Disorder: A randomized trial of schema focused therapy versus transference focused therapy. *Archives of General Psychiatry*, 63, 649–658.
Greenberg, L. (2008). Emotion and cognition in psychotherapy: The transforming power of affect. *Canadian Psychology*, 49, 49–59.
Hayes, S. C., Strosahl, K. D., & Wilson, K. G. (1999). *Acceptance and commitment therapy*. New York: Guilford.
Luoma, J. B., Hayes, S. C., & Walser, R. D. (2007). *Learn ACT: An Acceptance and Commitment Therapy skills training manual for therapists*. Oakland, CA: New Harbinger Publications.
Young, J. E., Klosko, J. S., Weishaar, M. E. (2003). *Schema therapy: A practitioner's guide*. New York: Guilford.

60 RESEARCH ON COUNSELING AND PSYCHOTHERAPY

Research in counseling focuses on process and outcome questions. The basic outcome question is this: Is counseling effective?

On the basis of previous research, most reviewers of the literature would say yes. But perhaps the best result of all the research conducted over the past five decades is that we are continually asking better questions (such as, What client characteristics interact with what type of treatment to produce what outcomes?), and we are continually improving the sophistication of the research methods employed to seek answers to these questions. Unfortunately, one might also argue that improved research sophistication is the most detectable outcome of past decades of counseling research.

That is, relatively few counselors pay attention to research beyond their training years, partially because research results often appear to add little to current practice. Research, theory, and practice, however, must be increasingly integrated if counseling is to progress as a science and benefit from the accompanying increase in knowledge and credibility.

Two attempts to integrate research into practice center on empirically validated or supported treatments (EVTs or ESTs) and treatment manuals. ESTs are specific counseling approaches that have been shown to produce effective outcomes for particular client problems (American Psychiatric Association, 2004; APA Task Force on Promotion and Dissemination of Psychological Procedures, 1995; Chambless, 1998). For example, one review of the literature indicates that cognitive therapy helps to alleviate depression when compared to control groups (Dobson, 1989). A number of published reviews (for example, Kendall, 1998; Pikoff, 1996) can provide information about specific ESTs. Perhaps the major strength *and* limitation of ESTs is that they are based on current research: Although preferable to methods with no research support, EST studies typically find that different counseling approaches outperform control groups but not each other (Wampold et al., 1997). EST studies also tend to ignore gender and cultural differences (see McGuire, 1999).

Treatment manuals are written materials meant to operationalize the specific methods employed in a particular counseling approach (Nathan, 1998). Often based on ESTs, treatment manuals provide a standardized approach to counseling and provide very specific guidance for counselor training. Numerous authors and publishers (for example, Linehan, 1993; Luborsky, 1993; Graywind Publication's Psychosocial Therapeutic Systems; The Psychological Corporation's TherapyWorks series; John Wiley's Practice Planner Series) provide treatment manuals for different approaches and problems.

Just as with multicultural counseling and clients of color, caution should be employed when applying ESTs to individual clients. Researchers conducting psychotherapy research typically operate from a nomothetic perspective that assumes that universal laws or principles govern human behavior. Nevertheless, one truism of counseling and psychotherapy research is that individuals do not respond uniformly to any type of intervention. Even with counseling approaches found to be effective, on average, for a particular problem or group of clients, some individuals will have shown no change or deterioration. While a research-based approach is a good starting point when planning for counseling with any particular client, counselors typically adopt an idiographic perspective. That is, they individualize their approach (recall Chapter 1), knowing that for any particular client, unique elements of that person's history, personality, and environment are likely to influence the success or failure of counseling.

References

APA Task Force on Promotion and Dissemination of Psychological Procedures. (1995). Training in and dissemination of empirically validated psychological treatments: Report and recommendation. *The Clinical Psychologist, 48,* 3–23.

Dawes, R. M. (1996). House of cards: Psychology and psychotherapy built on myth. New York: The Free Press.

Dobson, K. (1989). A meta-analysis of the efficacy of cognitive therapy for depression. *Journal of Consulting and Clinical Psychology, 57,* 414–419.

Eysenck, H. (1952). The effects of psychotherapy: An evaluation. *Journal of Consulting Psychology, 16,* 319–324.

Garfield, S., & Bergin, A. (1994). Handbook of psychotherapy and behavior change: An empirical analysis (4th ed.). New York: Wiley.

Gelso, C. (1979). Research in counseling: Methodological and professional issues. *The Counseling Psychologist, 8,* 7–35.

Kendall, P. C. (1998). Empirically supported psychological therapies. *Journal of Consulting and Clinical Psychology, 66,* 3–6.

Linehan, M. M. (1993). Cognitive-behavioral treatment of borderline personality disorder. New York: Guilford.

Luborsky, L. (1993). Recommendations for training therapists based on manuals for psychotherapy research. *Psychotherapy, 30,* 578–586.

McGuire, P. A. (1999). Multicultural summit cheers packed house. *APA Monitor, 30,* 26.

Nathan, P. E. (1998). Practice guidelines: Not yet ideal. *American Psychologist, 53,* 290–299.

Nathan, P. E. (2000). The Boulder model. *American Psychologist, 55,* 250–251.

Pikoff, H. B. (1996). *Treatment effectiveness handbook.* Buffalo, NY: Data for Decisions.

Smith, M., & Glass, G. (1977). Meta-analysis of psychotherapy outcome studies. *American Psychologist, 32,* 752–760.

Wampold, B. E., Mondin, G. W., Moody, M., Stich, F., Benson, K., & Ahn, H. (1997). A meta-analysis of outcome studies comparing bona fide psychotherapies: Empirically, "All Must Have Prizes." *Psychological Bulletin, 122,* 203–215.

Wilson, G. T. (1998). Manual-based treatment and clinical practice. *Clinical Psychology—Science and Practice, 5,* 363–375.

61 OTHER IMPORTANT SOURCES

Not all important publications fit neatly into the categories already listed. The following sources are also worthy of your attention.

REFERENCES

Adler, A. (1999). *The practice and theory of individual psychology* (2nd ed.). Totowa, NJ: Littlefield.

American Psychiatric Association. (2000). *Diagnostic and statistical manual of mental disorders* (Rev. 4th ed.). Washington, DC: Author.

Galanter, M., Castaneda, R., & Franco, H. (1998). Group therapy, self-help groups, and network therapy. In R. J. Frances & S. I. Miller (Eds.), *Clinical textbook of addictive disorders* (2nd ed., pp. 521–546). New York: Guilford.

Glasser, W. (1975). *Reality therapy.* New York: Harper & Row.

Lazarus, A. (1989). *The practice of multimodal therapy.* New York: McGraw-Hill.

Lerner, H. G. (2005). The dance of anger: A woman's guide to changing the patterns of intimate relationships. New York: Harper Paperbacks.

Zeig, J. (1997). The evolution of psychotherapy: The third conference. New York: Brunner/Mazel.

DISCUSSION QUESTIONS

To help you understand and apply the information in this chapter, consider these questions:

How do the approaches described in this chapter differ?

On what dimensions do the counseling approaches overlap?

Do you have preferences for the types of counseling approaches you feel most comfortable employing in counseling?

Beyond graduate school, how might you receive additional training in different or new counseling approaches?

How might you stay informed about research describing and evaluating counseling approaches?

REFERENCES

Abramowitz, J. S. (2002). Treatment of obsessive thoughts and cognitive rituals using exposure and response prevention. *Clinical Case Studies, 1*, 6–24.

Abramowitz, S., Weitz, L., Schwartz, J., Amura, S., Gomes, B., & Abramowitz, C. (1975). Comparative counselor inferences toward women with medical school aspirations. *Journal of College Student Personnel, 16*, 128–130.

Adams, J. F. (1997). Questions as interventions in therapeutic conversation. *Journal of Family Psychotherapy, 8*, 17–35.

Adams, J. F., Piercy, F. P., & Jurich, J. A. (1991). Effects of solution focused therapy's "Formula First Session Task" on compliance and outcome in family therapy. *Journal of Marital and Family Therapy, 17*, 277–290.

Allumbaugh, D. L., & Hoyt, W. T. (1999). Effectiveness of grief therapy: A meta-analysis. *Journal of Counseling Psychology, 46*, 370–380.

American Counseling Association. (2005). *ACA code of ethics*. Alexandria, VA. Retrieved March 14, 2006, from http://www.counseling.org/Resources/CodeOfEthics/TP/Home/CT2.aspx.

American Psychiatric Association. (2000). *Diagnostic and statistical manual of mental disorders* (Rev. 4th ed.). Washington, DC: Author.

American Psychiatric Association. (2004). *Practice guidelines for treatment of psychiatric disorders: Compendium 2004*. Washington, DC: American Psychiatric Press.

American Psychological Association. (2002). *Ethical principles of psychologists and code of conduct*. Retrieved September 25, 2003, from www.apa.org/ethics.

American Rehabilitation Counseling Association. (2002). *Code of professional ethics for rehabilitation counselors*. Retrieved March 14, 2006, from www.arcaweb.org/pdf/code_ethics_2002.pdf.

American School Counselor Association. (2004). *Ethical standards for school counselors*. Retrieved March 14, 2006, from www.schoolcounselor.org/files/ethical%20standards.pdf.

Amir, N., Stafford, J., Freshman, M. S., & Fora, E. B. (1998). Relationship between trauma narratives and trauma pathology. *Journal of Traumatic Stress, 11*, 385–392.

Anderson, T. (2004). "To tell my story": Configuring interpersonal relations within narrative process. In L. E. Angus & J. McLeod (Eds.), *The handbook of narrative and psychotherapy* (pp. 315–329). Thousand Oaks, CA: Sage.

Angus, L. E., & McLeod, J. (2004). Toward an integrative framework for understanding the role of narrative in the psychotherapy process. In L. E. Angus & J. McLeod (Eds.), *The handbook of narrative and psychotherapy* (pp. 367–374). Thousand Oaks, CA: Sage.

Anthony, S., & Pagano, G. (1998). The therapeutic potential for growth during the termination process. *Clinical Social Work Journal, 26*, 218–296.

APA Task Force on Promotion and Dissemination of Psychological Procedures. (1995). Training in and dissemination of empirically validated psychological treatments: Report and recommendation. *The Clinical Psychologist, 48*, 3–23.

Araoz, D. L., & Carrese, M. A. (1996). *Solution-oriented brief therapy for adjustment disorders: A guide for providers under managed care*. New York: Brunner/Mazel.

Association for Specialists in Group Work. (1989). *Ethical guidelines for group counselors*. Alexandria, VA: American Counseling Association.

Ballou, M. (1995). Assertiveness training. In M. Ballou (Ed.), *Psychological interventions: A guide to strategies* (pp. 125–135). Westport, CT: Praeger.

Bandler, R., & Grinder, J. (1975). *The structure of magic*. Palo Alto, CA: Science and Behavior Books.

Bandura, A. (1977). Self-efficacy theory: Toward a unifying view of behavioral change. *Psychological Review, 84*, 191–215.

Bandura, A. (1997). *Self-efficacy: The exercise of control*. New York: Freeman.

Barber, J. P. (1994). Efficacy of short-term dynamic psychotherapy: Past, present, and future. *Journal of Psychotherapy Practice and Research, 3*, 108–121.

Barrett, L. B. (2006). Solving the motion paradox: Categorization and the experience of emotion. 2006 *Personality & Social Psychology Review, 10*, 20–46.

Barrett-Lennard, G. (1974). Experiential learning groups. *Psychotherapy: Theory, Research, and Practice, 11*, 71–75.

Bednar, R. L., & Kaul, T. J. (1994). Experiential group research: Can the canon fire? In A. E. Bergin & S. L. Garfield (Eds.), *Handbook of psychotherapy and behavior change* (4th ed., pp. 631–663). New York: Wiley.

Bergin, A. E., & Garfield, S. L. (1994). Overview, trends, and future issues. In A. E. Bergin & S. L. Garfield (Eds.), *Handbook of psychotherapy and behavior change* (4th ed., pp. 190–228). New York: Wiley.

Bersoff, D. N. (2003). *Ethical conflicts in psychology* (3rd ed.). Washington, DC: American Psychological Association.

Betz, N., & Fitzgerald, L. (1993). Individuality and diversity: Theory and research in counseling psychology. *Annual Review of Psychology, 44*, 343–381.

Beutler, L. E. (1984). Comparative effects of group psychotherapies in a short-term inpatient setting: An experience with deterioration effects. *Psychiatry, 47*, 66–76.

Beutler, L. E., & Harwood, T. M. (2000). *Prescriptive psychotherapy: A practical guide to systematic treatment selection*. Oxford: Oxford University Press.

Borders, L. D., Bloss, K. K., Cashwell, C. S., & Rainey, L. M. (1994). Helping students apply the scientist-practitioner model: A teaching approach. *Counselor Education and Supervision, 34*, 172–179.

Bowen, M. (1978). *Family therapy in clinical practice*. New York: Aronson.

Bowman, R. L., & Bowman, V. E. (1998). Life on the electronic frontier: The application of technology to group work. *Journal for Specialists in Group Work, 23*, 428–445.

Brady, J. L., Healy, F. C., Norcross, J. C., & Guy, J. D. (1995). Stress in counsellors: An integrative research review. In W. Dryden (Ed.), *The stresses of counselling in action* (pp. 1–27). London: Sage.

Brammer, L., & Shostrom, E. (1989). *Therapeutic psychology* (5th ed.). Englewood Cliffs, NJ: Prentice-Hall.

Broverman, L., Broverman, D., Clarkson, E., Rosenkrantz, P., & Vogel, S. (1970). Sex-role stereotypes and clinical judgments of mental health. *Journal of Consulting and Clinical Psychology, 34*, 1–7.

Brown, L. (1994). *Subversive dialogues: Theory in feminist therapy*. New York: Basic Books.

Brown, L. (2000). Feminist therapy. In C. R. Snyder & R. E. Ingram (Eds.), *Handbook of psychological change: Psychotherapy processes and practices for the 21st century*. New York: Wiley.

Bruner, J. (2004). The narrative creation of self. In L. E. Angus & J. McLeod (Eds.), *The handbook of narrative and psychotherapy* (pp. 3–14). Thousand Oaks, CA: Sage.

Budman, S. H. (1983). Brief therapy in the year 2000 and beyond: Looking back while looking forward. In S. H. Budman (Ed.), *Forms of brief therapy* (pp. 462–470). New York: Guilford.

Butcher, J., & Koss, M. (1978). Research on brief and crisis-oriented psychotherapies. In S. Garfield & A. Bergin (Eds.), *Handbook of psychotherapy and behavior change: An empirical analysis* (2nd ed., pp. 725–768). New York: Wiley.

Campbell-Sills, L., & Barlow, D. H. (2007). Incorporating emotion regulation into conceptualizations and treatments of anxiety and mood disorders. In J. J. Gross (Ed.), *Handbook of emotion regulation* (pp. 542–559). New York: Guilford.

Canadian Psychological Association. (2000). *Canadian code of ethics for psychologists* (3rd ed.). Ottawa, Ontario. Retrieved March 15, 2006, from www.pre.ethics.gc.ca/english/pdf/links/Canadian%20Code%20of%20Ethics%20for%20Psychologists%20_2000.pdf.

Caplan, G. (1961). *An approach to community mental health*. New York: Grune & Stratton.

Carter, R. T. (1991). Cultural values: A review of empirical research and implications for counseling. *Journal of Counseling and Development, 70*, 164–173.

Casas, J. M. (1984). Policy, training, and research in counseling psychology: The racial/ethnic minority perspective. In S. Brown & R. Lent (Eds.), *Handbook of counseling psychology* (pp. 785–831). New York: Wiley.

Cashdan, S. (1988). *Object relations therapy.* New York: Norton.

Chambless, D. (1998). Defining empirically supported therapies. *Journal of Counseling and Clinical Psychology, 66,* 7–18.

Christensen, A., & Jacobson, N. S. (1994). Who (or what) can do psychotherapy: The status and challenge of nonprofessional therapies. *Psychological Science, 5,* 8–14.

Claiborn, C. D., & Goodyear, R. K. (2005). Feedback in therapy. *Journal of Clinical Psychology, 61,* 209–217.

Claiborn, C., & Hanson, W. (1999). Test interpretation: A social influence perspective. In J. W. Lichtenberg & R. K. Goodyear (Eds.), *Scientist-practitioner perspectives on test interpretation* (pp. 151–166). Boston: Allyn & Bacon.

Close, H. T. (1998). *Metaphor in psychotherapy: Clinical applications of stories and allegories.* San Luis Obispo, CA: Impact.

Clunis, D. M., & Green, G. D. (2005). *Lesbian couples* (4th ed). Seattle, WA: Seal Press.

Coleman, H. L., Wampold, B. E., & Casali, S. L. (1995). Ethnic minorities' ratings of ethnically similar and European American counselors: A meta-analysis. *Journal of Counseling Psychology, 42,* 55–64.

Corey, G. (1995). *Theory and practice of group counseling* (4th ed.). Pacific Grove, CA: Brooks/Cole.

Cormican, J. (1978, March). Linguistic issues in interviewing. *Social Casework,* 145–151.

Cormier, W., & Cormier, L. (1998). *Interviewing strategies for helpers: Fundamental skills and cognitive behavioral intentions* (4th ed.). Pacific Grove, CA: Brooks/Cole.

Crits-Christoph, P. (1992). The efficacy of brief dynamic psychotherapy: A meta-analysis. *American Journal of Psychiatry, 149,* 151–158.

Cummings, A. L., Hallberg, E. T., Slemon, A., & Martin, J. (1992). Participants' memories for therapeutic events and ratings of session effectiveness. *Journal of Cognitive Psychotherapy: An International Quarterly, 6,* 113–124.

Davis, S. R., & Meier, S. T. (2000). *The elements of managed care: A guide for helping professionals.* Pacific Grove, CA: Brooks/ Cole.

de Luynes, M. (1995). Neurolinguistic programming. *Educational and Child Psychology, 12,* 34–47.

deShazer, S. (1982). *Patterns of brief family therapy.* New York: Guilford.

DeSpelder, L. A., & Strickland, A. L. (2004). *The last dance: Encountering death and dying.* New York: McGraw-Hill.

Dobson, K. (1989). A meta-analysis of the efficacy of cognitive therapy for depression. *Journal of Consulting and Clinical Psychology, 57,* 414–419.

Does therapy help? (1995, November). *Consumer Reports, 60,* 734–739.

Eells, T. D. (Ed.). (1997). *Handbook of psychotherapy case formulation.* New York: Guilford.

Egan, G. (2001). *The skilled helper: A problem-management and opportunity—development approach to helping* (7th ed.). Pacific Grove, CA: Brooks/Cole.

Eisler, R. (1976). Assertive training in the work situation. In J. Krumboltz & C. Thoresen (Eds.), *Counseling methods* (pp. 29–36). New York: Holt, Rinehart & Winston.

Ellis, A. (1998). *Rational emotive behavior therapy: A therapist's guide.* San Luis Obispo, CA: Impact.

Ellis, A., & Grieger, R. (1977). *Handbook of rational-emotive therapy.* New York: Springer.

Ellis, A., & Harper, R. (1976). *A new guide to rational living.* North Hollywood, CA: Wilshire.

Engels, D. W. (2004). *The professional counselor: Portfolio, competencies, performance guidelines, and assessment* (3rd ed.). Alexandria, VA: American Counseling Association.

Enns, C. (1997). *Feminist theories and feminist psychotherapies.* New York: Haworth.

Erhard & Erhard-Weiss, (2007). The emergence of counseling in traditional cultures: Ultra-Orthodox Jewish and Arab communities in Israel. *International Journal for the Advancement of Counselling, 29,* 149–158.

Fischer, A. R., Jome, L. M., & Atkinson, D. R. (1998). Reconceptualizing multicultural counseling: Universal healing conditions in a culturally specific context. *The Counseling Psychologist, 26,* 525–588.

Flaskerud, J. H. (1990). Matching client and therapist ethnicity, language, and gender: A review of research. *Issues in Mental Health Nursing, 11,* 321–336.

Foa, E. B., Hembree, E. A., & Rothbaum, B. O. (2007). Prolonged exposure therapy for PTSD: Emotional processing of traumatic experiences. Oxford: Oxford University Press.

Foa, E. B., Molnar, C., & Cashman, L. (1995). Change in rape narratives during exposure therapy for posttraumatic stress disorder. *Journal of Traumatic Stress—Special Research on Traumatic Memory, 8,* 675–690.

Fortner, B. V. (1999). The effectiveness of grief counseling and therapy: A quantitative review (Doctoral dissertation, University of Memphis). *Dissertation Abstracts International, 60,* 4221.

Fouad, N. A., & Chan, P. M. (1999). Gender and ethnicity: Influence on test interpretation and reception. In J. W. Lichtenberg & R. K. Goodyear (Eds.), *Scientist-practitioner perspectives on test interpretation* (pp. 31–58). Boston: Allyn & Bacon.

Frank, E. (1991). Interpersonal psychotherapy as a maintenance treatment for patients with recurrent depression. *Psychotherapy, 28,* 259–266.

Frank, J. (1971). Therapeutic factors in psychotherapy. *American Journal of Psychotherapy, 25,* 350–361.

Frank, J. D., & Frank, J. B. (1991). *Persuasion and healing: A comparative study of psychotherapy* (3rd ed.). Baltimore: Johns Hopkins University Press.

Freudenberger, H. (1974). Staff burnout. *Journal of Social Issues, 30,* 159–165.

Fridja, N. H., & Sundararajan, L. (2007). Emotion refinement. *Perspectives on Psychological Science, 2,* 227–241.

Froyd, J. E., Lambert, M. J., & Froyd, J. D. (1996). A review of practices of psychotherapy outcome measurement. *Journal of Mental Health, 5,* 11–15.

Fuhriman, A., & Burlingame, G. M. (Eds.). (1994). *Handbook of group psychotherapy.* New York: Wiley.

Garfield, S. (1994). Research on client variables in psychotherapy. In A. E. Bergin & S. L. Garfield (Eds.), *Handbook of psychotherapy and behavior change* (4th ed., pp. 190–228). New York: Wiley.

Gelso, C. (1979). Research in counseling: Methodological and professional issues. *The Counseling Psychologist, 8,* 7–35.

Glidden-Tracey, C. E., & Wagner, L. (1995). Gender salient attribute × treatment interaction effects on ratings of two analogue counselors. *Journal of Counseling Psychology, 42,* 223–231.

Goldfried, M. R. (1983). The behavior therapist in clinical practice. *Behavior Therapist, 6,* 45–46.

Goldman, L. (1976). A revolution in counseling research. *Journal of Counseling Psychology, 23,* 543–552.

Gray, B. (1991). *The profit motive and patient care.* Cambridge, MA: Harvard University Press.

Greenberg, L. (2008). Emotion and cognition in psychotherapy: The transforming power of affect. *Canadian Psychology, 49,* 49–59.

Greenberg, L. S., & Angus, L. E. (2004). The contributions of emotion processes to narrative chance in psychotherapy: A dialectical constructivist approach. In L. E. Angus & J. McLeod (Eds.), *The handbook of narrative and psychotherapy* (pp. 331–364). Thousand Oaks, CA: Sage.

Greenson, R. (1965). The working alliance and the transference neurosis. *Psychoanalytic Quarterly, 34,* 155–181.

Gunzburger, D., Henggeler, S., & Watson, S. (1985). Factors related to premature termination of counseling relationships. *Journal of College Student Personnel, 26,* 456–460.

Gurman, A., & Kniskern, D. (1978). Research on marital and family therapy: Progress, perspective, and prospect. In S. Garfield & A. Bergin (Eds.), *Handbook of psychotherapy and behavior change: An empirical analysis* (2nd ed., pp. 817–902). New York: Wiley.

Hall, L. K. (2008). *Counseling military families.* New York: Routledge.

Hansen, J., Rossberg, R. R., & Cramer, S. H. (1994). *Counseling: Theory and process* (4th ed.). Boston: Allyn & Bacon.

Hardtke, K. K., & Angus, L. E. (2004). The narrative assessment interview. In L. E. Angus & J. McLeod (Eds.), *The handbook of narrative and psychotherapy* (pp. 247–262). Thousand Oaks, CA: Sage.

Hartman, D. P. (1984). Assessment strategies. In D. H. Barlow & M. Hersen (Eds.), *Single case environmental designs* (pp. 107–129). New York: Pergamon Press.

Haynes, S., Leisen, M. B., & Blaine, D. (1997). Design of individualized behavioral treatment programs using functional analytic clinical case models. *Psychological Assessment, 9,* 334–348.

Hayes, S. C., Strosahl, K. D., & Wilson, K. G. (1999). *Acceptance and commitment therapy.* New York: Guilford.

Heinlen, K. T., Welfel, E. R., Richmond, E. N., & Rak, C. F. (2003). The scope of WebCounseling: A survey of services and compliance with NBCC Standards for the Ethical Practice of WebCounseling. *Journal of Counseling & Development, 81,* 61–69.

Helms, J. (1979). Perceptions of a sex-fair counselor and her client. *Journal of Counseling Psychology, 26,* 504–513.

Helms, J., & Cook, D. (1999). *Using race and culture in counseling and psychotherapy: Theory and process.* Needham, MA: Allyn & Bacon.

Heppner, P. P., Kivlighan, D. M., Jr., & Wampold, B. E. (1999). *Research design in counseling* (2nd ed.). Pacific Grove, CA: Brooks/Cole.

Herlihy, B., & Corey, G. (Eds.). (2005). *ACA ethical standards casebook* (6th ed.). Alexandria, VA: American Counseling Association.

Herr, E., & Best, P. (1984). Computer technology and counseling: The role of the profession. *Journal of Counseling and Development, 63,* 192–195.

Hill, M., & Ballou, M. (1998). Making feminist therapy: A practice survey. *Women and Therapy, 21,* 1–16.

Hodgson, R., & Rachman, S. (1974). Desynchrony in measures of fear. *Behaviour Research and Therapy, 12,* 319–326.

Hoehn-Saric, R., Frank, I., Imber, S., Nash, E., Stone, A., & Battle, C. (1964). Systematic preparation of patients for psychotherapy. I. Effects on therapy behavior and outcome. *Journal of Psychiatric Research, 2,* 267–281.

Holloway, E. (1995). *Clinical supervision.* Thousand Oaks, CA: Sage.

Hummel, T. J. (1999). The usefulness of tests in clinical decisions. In J. W. Lichtenberg & R. K. Goodyear (Eds.), *Scientist-practitioner perspectives on test interpretation* (pp. 59–112). Needham Heights, MA: Allyn & Bacon.

Ivey, A. (1980). Counseling 2000: Time to take charge! *The Counseling Psychologist, 8,* 12–16.

Ivey, A. (2002). *Intentional interviewing and counseling* (4th ed.). Pacific Grove, CA: Brooks/Cole.

Izard, C. E. (2007). Basic emotions, natural kinds, emotion schemas, and a new paradigm. *Perspectives on Psychological Science, 2,* 260–280.

Jacobs, D. G. (Ed.). (1999). *The Harvard Medical School guide to suicide assessment and intervention.* San Francisco: Jossey-Bass.

Jenkins, S. R., & Maslach, C. (1994). Psychological health and involvement in interpersonally demanding occupations: A longitudinal perspective. *Journal of Organizational Behavior, 15,* 101–127.

Jennings, L., & Skovholt, T. M. (1999). The cognitive, emotional, and relational characteristics of master therapists. *Journal of Counseling Psychology, 46,* 3–11.

Jordan, J. R., & Neimeyer, R. A. (2003). Does grief counseling work? *Death Studies, 27,* 765–786.

Jourard, S. (1971). *Self-disclosure: An elemental analysis of the transparent self.* New York: Wiley.

Juang, S., & Tucker, C. M. (1991). Factors in marital adjustment and their interrelationships: A comparison of Taiwanese couples in America and Caucasian American couples. *Journal of Multicultural Counseling and Development, 19,* 22–31.

Kahn, J., & Scott, N. (1998). Predictors of research productivity and science-related career goals among counseling psychology doctoral students. *The Counseling Psychologist, 25,* 38–67.

Kahn, M. (1997). *Between therapist and client* (Rev. ed.). New York: Freeman.

Kane, A. S., & Tryon, G. S. (1988). Predictors of premature termination from counseling at semester recess. *Journal of College Student Development, 29,* 562–563.

Kassan, L. D. (1996). *Shrink rap: Sixty psychotherapists discuss their work, their lives, and the state of their field.* Northvale, NJ: Aronson.

Kaul, T. J., & Bednar, R. L. (1994). Pretraining and structure: Parallel lines yet to meet. In A. Fuhriman & G. M. Burlingame (Eds.), *Handbook of group psychotherapy* (pp.155–188). New York: Wiley.

Kazdin, A. E. (1994). Psychotherapy for children and adolescents. In A. E. Bergin & S. L. Garfield (Eds.), *Handbook of psychotherapy and behavior change* (4th ed., pp. 543–594). New York: Wiley.

Kazdin, A. E. (2000). *Psychotherapy for children and adolescents: Directions for research and practice.* New York: Oxford University Press.

Kees, N. L. (2005). Women's voices, women's lives: An introduction to the special issue on women and counseling. *Journal of Counseling and Development, 83,* 259–261.

Kelly, M. A. R., Roberts, J. E., & Ciesla, J. A. (2005). Sudden gains in cognitive behavioral treatment for depression: When do they occur and do they matter? *Behavior Research and Therapy, 43,* 703–714.

Kendall, P. C. (1998). Empirically supported psychological therapies. *Journal of Consulting and Clinical Psychology, 66,* 3–6.

Kendall, P. C., Kipnis, D., & Otto-Salaj, L. (1992). When clients don't progress: Influences on and explanations of therapeutic progress. *Cognitive Therapy and Research, 16,* 269–281.

Kenyon, P. (1998). *What would you do? An ethical case workbook for human service professionals.* Pacific Grove, CA: Brooks/Cole.

Kessler, K. A. (1998). History of managed behavioral health care and speculations about its future. *Harvard Review of Psychiatry, 6,* 155–159.

Kiesler, D. J. (1971). Experimental designs in psychotherapy research. In A. Bergin & S. Garfield (Eds.), *Handbook of psychotherapy and behavior change* (pp. 36–74). New York: Wiley.

Kilburg, R., Nathan, P., & Thoreson, R. (1986). *Professionals in distress.* Washington, DC: American Psychological Association.

Kirschner, T., Hoffman, M., & Hill, C. E. (1994). Case study of the process and outcome of career counseling. *Journal of Counseling Psychology, 41,* 216–226.

Koocher, G. P., & Pollin, I. (1994). Medical crisis counseling: A new service delivery model. *Journal of Clinical Psychology in Medical Settings, 1,* 291–299.

Kopta, S. M., Howard, K. I., Lowry, J. L., & Beutler, L. E. (1994). Patterns of symptomatic recovery in psychotherapy. *Journal of Consulting and Clinical Psychology, 62,* 1009–1016.

Kottler, J. A. (Ed.). (1997). *Finding your way as a counselor.* Alexandria, VA: American Counseling Association.

Krause, M. S., Howard, K. I., & Lutz, W. (1998). Exploring individual change. *Journal of Consulting and Clinical Psychology, 66,* 838–845.

Kremer, T. G., & Gesten, E. L. (1998). Confidentiality limits of managed care and clients' willingness to self-disclose. *Professional Psychology—Research and Practice, 29*, 553–558.

Krumboltz, J. (Ed.). (1966). *Revolution in counseling.* Boston: Houghton Mifflin.

Lambert, M. J. (2005). Early response in psychotherapy: Further evidence for the importance of common factors rather than "placebo effects." *Journal of Clinical Psychology, 61*, 855–869.

Lambert, M. J., & Bergin, A. E. (1994). The effectiveness of psychotherapy. In A. E. Bergin & S. L. Garfield (Eds.), *Handbook of psychotherapy and behavior change* (4th ed., pp. 143–189). New York: Wiley.

Lambert, M. J., & Cattani-Thompson, K. (1996). Current findings regarding the effectiveness of counseling: Implications for practice. *Journal of Counseling and Development, 74*, 601–608.

Lambert, M. J, Whipple, J. L., Smart, D. W., Vermeesch, D. A., Nielsen, S. L., & Hawkins, E. J. (2001). The effects of providing therapists with feedback on patient progress during psychotherapy: Are outcomes enhanced? *Psychotherapy Research, 11*, 49–68.

Landman, J., & Dawes, R. (1982). Psychotherapy outcome. *American Psychologist, 37*, 504–516.

Langs, R. (1973). *The techniques of psychoanalytic psychotherapy (Vol. 1).* New York: Aronson.

Leong, F. T. L., & Blustein, D. L. (2000). Toward a global vision of counseling psychology. *The Counseling Psychologist, 28*, 5–9.

Leong, R. T. L., & Ponterotto, J. G. (2003). A proposal for international counseling psychology in the United States: Rationales, recommendations, and challenges. *The Counseling Psychologist, 31*, 381–395.

Leung, S. A. (2003). A journey worth traveling: Globalization of Counseling Psychology. *The Counseling Psychologist, 31*, 419.

Levitt, H. (2002). Voicing the unvoiced: Narrative formulation and silences. *Journal of Counseling Psychology Quarterly, 15*, 333–350.

Levitt, H. M., & Rennie, D. L. (2004). Narrative activity. In L. E. Angus & J. McLeod (Eds.), *The handbook of narrative and psychotherapy* (pp. 299–313). Thousand Oaks, CA: Sage.

Lichtenberg, J. W., & Goodyear, R. K. (1999). *Scientist-practitioner perspectives on test interpretation.* Boston: Allyn & Bacon.

Lieberman, M., Yalom, L., & Miles, M. (1973). *Encounter groups: First facts.* New York: Basic Books.

Lijtmaer, R. M. (1998). Psychotherapy with Latina women. *Feminism and Psychology, 8*, 538–543.

Linehan, M. M. (1993). *Cognitive-behavioral treatment of borderline personality disorder.* New York: Guilford.

Lopez, S. R. (1989). Patient variable biases in clinical judgment: Conceptual overview and methodological considerations. *Psychological Bulletin, 106*, 184–203.

Lott, B., & Rocchio, L. M. (1998). Standing up, talking back, and taking charge: Strategies and outcomes in collective action against sexual harassment. In L. H. Collins & J. C. Chrisler (Eds.), *Career strategies for women in academe: Arming Athena* (pp. 249–273). Thousand Oaks, CA: Sage.

Luborsky, L. (1993). Recommendations for training therapists based on manuals for psychotherapy research. *Psychotherapy, 30*, 578–586.

Luoma, J. B., Hayes, S. C., & Walser, R. D. (2007). *Learn ACT: An Acceptance Commitment Therapy skills training manual for therapists.* Oakland, CA: New Harbinger Publications.

MacKinnon, R., & Michels, R. (1971). *The psychiatric interview in clinical practice.* Philadelphia: Saunders.

Mahoney, M. (1987, August 31). *Plasticity and power: Emerging emphases in theories of human change.* Paper presented at the 95th annual meeting of the American Psychological Association, New York.

Marshall, J. (2006). Counseling on the front line: Providing a safe refuge for military personnel to discuss emotional wounds. *Counseling Today, 48*, 32–33.

Martin, J. (1990). Individual differences in client reactions to counselling and psychotherapy: A challenge for research. *Counselling Psychology Quarterly, 3*, 67–83.

Martin, J. (1994). *The construction and understanding of psychotherapeutic change: Conversations, memories, and theories.* New York: Teachers College Press.

Mash, E. J., & Hunsley, J. (1993). Assessment considerations in the identification of failing psychotherapy: Bringing the negatives out of the darkroom. *Psychological Assessment, 5*, 292–301.

Maslach, C., & Leiter, M. P. (1997). *The truth about burnout: How organizations cause personal stress and what to do about it.* San Francisco: Jossey-Bass.

Mayerson, N. (1984). Preparing clients for group therapy: A critical review and theoretical formulation. *Clinical Psychology Review, 4*, 191–213.

McCarthy, P. (1982). Differential effects of counselor self-referent responses and counselor status. *Journal of Counseling Psychology, 29*, 125–131.

McCarthy, P., & Betz, N. (1978). Differential effects of self-disclosing versus self—involving counselor statements. *Journal of Counseling Psychology, 25*, 251–256.

McGuire, P. A. (1999). Multicultural summit cheers packed house. *APA Monitor, 30*, 26.

Meehl, P. (1956). Wanted—A good cookbook. *American Psychologist, 11,* 262–272.

Meier, S. T. (1986). Stories about counselors and computers: Their use in workshops. *Journal of Counseling and Development, 65,* 100–103.

Meier, S. T. (1987). An unconnected special issue. *American Psychologist, 42,* 881.

Meier, S. T. (1999). Training the practitioner-scientist: Bridging case conceptualization, assessment, and intervention. *The Counseling Psychologist, 27,* 589–613.

Meier, S. T. (2001). Investigating clinical trainee development through item analysis of self-reported skills: The identification of perceived credibility. *The Clinical Supervisor, 20,* 25–38.

Meier, S. T. (2003). *Bridging case conceptualization, assessment, and intervention.* Thousand Oaks, CA: Sage.

Meier, S. T. (2008). Measuring change in counseling and psychotherapy. New York: Guilford.

Meier, S. T., & Schwartz, E. (2007). *Negative changes on new outcome assessments with adolescent clients: A social desirability effect?* Unpublished manuscript, University at Buffalo.

Meier, S. T., & Vermeersch, D. (2007). *What changes in counseling and psychotherapy?* Unpublished manuscript, University at Buffalo.

Merten, J., Anstadt, T., Ullrich, R., Krause, R., & Buchheim, P. (1996). Emotional experience and facial behavior during the psychotherapeutic process and its relation to treatment outcome: A pilot study. *Psychotherapy Research, 6,* 198–212.

Minuchin, S. (1974). *Families and family therapy.* Cambridge, MA: Harvard University Press.

Minuchin, S., & Fishman, H. C. (2004). *Family therapy techniques.* Cambridge, MA: Harvard University Press.

Mohr, D. C. (1995). Negative outcome in psychotherapy: A critical review. *Clinical Psychology: Science and Practice, 2,* 1–27.

Monahan, J. (1995). Limiting therapist exposure to *Tarasoff* liability: Guidelines for risk containment. In D. Bersoff (Ed.), *Ethical conflicts in psychology* (pp. 174–181). Washington, DC: American Psychological Association.

Montague, J. (1996). Counseling families from diverse cultures: A nondeficit approach. *Journal of Multicultural Counseling and Development, 24,* 37–41.

Moses, E. B., & Barlow, D. B. (2006). A new unified treatment approach for emotional disorders based on emotion science. *Current Directions in Psychological Science, 15,* 146–150.

Moxley, D. (1989). *The practice of case management.* Newbury Park, CA: Sage.

Myers, J. E. (1998). Bibliotherapy and DCT: Co-constructing the therapeutic metaphor. *Journal of Counseling and Development, 76,* 243–250.

Myrick, R. D., & Sabella, R. A. (1995). Cyberspace: New place for counselor supervision. *Elementary School Guidance and Counseling, 30,* 35–44.

Nathan, P. E. (1998). Practice guidelines: Not yet ideal. *American Psychologist, 53,* 290–299.

National Association of School Psychologists. (1997). *Principles for professional ethics.* Retrieved March 15, 2006, from www.nasponline.org/certification/ethics.html.

National Board for Certified Counselors. (1989). *National Board for Certified Counselors: Code of ethics.* Alexandria, VA: Author.

National Board for Certified Counselors. (1997). *Standards for the ethical practice of WebCounseling.* Greensboro, NC: Author.

National Career Development Association. (2003). *National Career Development Association ethical standards.* Retrieved March 15, 2006, from www.ncda.org/pdf/EthicalStandards.pdf.

Ng, K. S. (1999). *Counseling Asian families from a systems perspective.* Alexandria, VA: American Counseling Association.

O'Connell, B. (2005). *Solution-focused therapy* (2nd ed.). London: Sage.

Odell, M., & Quinn, W. H. (1998). Therapist and client behaviors in the first interview: Effects on session impact and treatment duration. *Journal of Marriage and Family Counseling, 24,* 369–388.

Ogles, B. M., Lambert, M. J., & Masters, K. S. (1996). *Assessing outcome in clinical practice.* New York: Allyn & Bacon.

Orlinsky, D. E., Grawe, K., & Parks, B. K. (1994). Process and outcome in psychotherapy: Noch einmal. In A. E. Bergin & S. L. Garfield (Eds.), *Handbook of psychotherapy and behavior change* (4th ed., pp. 270–376). New York: Wiley.

Orlinsky, D., & Howard, K. (1978). The relation of process to outcome in psychotherapy. In S. Garfield & A. Bergin (Eds.), *Handbook of psychotherapy and behavior change: An empirical analysis* (2nd ed., pp. 283–330). New York: Wiley.

Paniagua, F. A. (2005). *Assessing and treating culturally diverse clients* (3rd ed.). Thousand Oaks, CA: Sage.

Parham, Thomas A. (Ed.). (2002). *Counseling persons of African descent: Raising the bar of practitioner competence.* Thousand Oaks, CA: Sage.

Paul, G. L. (Ed.). (1986). *Assessment in residential treatment settings.* Champaign, IL: Research Press.

Paul, G. L., & Menditto, A. A. (1992). Effectiveness of inpatient treatment programs for mentally ill adults in public psychiatric facilities. *Applied & Preventive Psychology, 1,* 41–63.

Pedersen, P. B., & Leong, F. T. L. (1997). Counseling in an international context. *The Counseling Psychologist, 25,* 117–122.

Pennebaker, J. W. & Beall, S. K. (1986). Confronting a traumatic event: Toward an understanding of inhibition and disease. *Journal of Abnormal Psychology, 95,* 274–281.

Pennebaker, J. W., Mehl, M. R., & Niederhoffer, K. G. (2003). Psychological aspects of natural language use: Our words, our selves. *Annual Review of Psychology, 54,* 547–577.

Pennebaker, J. W., Zech, E., & Rimé, B. (2001). Disclosing and sharing emotion: Psychological, social and health consequences. In M. S. Stroebe, R. O. Hansson, W. Stroebe, & H. Schut (Eds.), *Handbook of bereavement research: Consequences, coping, and care* (pp. 517–544). Washington, DC: American Psychological Association.

Perez, J. E. (1999). Clients deserve empirically supported treatments, not romanticism. *American Psychologist, 54,* 205–206.

Perlstein, M. (1998). Where, oh where, has the therapeutic alliance gone? Disquieting log-jams in the therapeutic relationship. *Women and Therapy, 21,* 63–68.

Petrocelli, J. V. (2002). Processes and stages of change: Counseling with the transtheoretical model of change. *Journal of Counseling & Development, 80,* 22–30.

Pikoff, H. B. (1996). *Treatment effectiveness handbook.* Buffalo, NY: Data for Decisions.

Polkinghorne, D. E. (2004). Narrative therapy and postmodernism. In L. E. Angus & J. McLeod (Eds.), *The handbook of narrative and psychotherapy* (pp. 53–67). Thousand Oaks, CA: Sage.

Pope, B. (1979). *The mental health interview.* New York: Pergamon Press.

Pope, K. (1988). How clients are harmed by sexual contact with mental health professionals: The syndrome and its prevalence. *Journal of Counseling and Development, 67,* 222–226.

Pope, K. S., & Tabachnick, B. G. (1994). Therapists as patients: A national survey of psychologists' experiences, problems, and beliefs. *Professional Psychology: Research and Practice, 25,* 247–258.

Pope, K. S., & Vasquez, M. J. T. (1998). *Ethics in psychotherapy and counseling: A practical guide* (2nd ed.). San Francisco: Jossey-Bass.

Prieto, L. R., & Scheel, K. R. (2002). Using case documentation to strengthen counselor trainees' case conceptualization skills. *Journal of Counseling & Development, 80,* 11–21.

Prochaska, J. O. (1995). An eclectic and integrative approach: Transtheoretical therapy. In A. S. Gurman & S. B. Messer (Eds.), *Essential psychotherapies* (pp. 403–440). New York: Oxford University Press.

Prochaska, J. O., & DiClemente, C. C. (1983). Stages and processes of self-change in smoking: Toward an integrative model of change. *Journal of Consulting and Clinical Psychology, 5,* 390–395.

Prochaska, J. O., Johnson, S., & Lee, P. (1998). The transtheoretical model of behavior change. In S. A. Shumaker & E. B. Schron (Eds.), *The handbook of health behavior change* (2nd ed., pp. 59–84). New York: Springer.

Prochaska, J. O., & Norcross, J. (2006). *Systems of psychotherapy: A transtheoretical analysis* (6th ed.). Pacific Grove, CA: Wadsworth.

Puryear, D. (1979). *Helping people in crisis: A practical family-oriented approach to effective crisis intervention.* San Francisco: Jossey-Bass.

Quick, E. K. (1996). *Doing what works in brief therapy: A strategic solution focused approach.* San Diego: Academic Press.

Rachman, S., & Hodgson, R. (1974). Synchrony and desynchrony in fear and avoidance. *Behavior Research and Therapy, 12,* 311–318.

Ramirez, M., III. (1999). *Multicultural psychotherapy: An approach to individual and cultural differences* (2nd ed.). Boston: Allyn & Bacon.

Rando. T. (1993). *Treatment of complicated mourning.* Champaign, IL: Research Press.

Rawlings, E. I., & Carter, D. K. (1977). Feminist and nonsexist psychotherapy. In E. I. Rawlings & D. K. Carter (Eds.), *Psychotherapy for women* (pp. 49–76). Springfield, IL: Thomas.

Reich, W. P. (1998). Metaphor dialog in psychotherapy. *American Journal of Clinical Hypnosis, 40,* 306–319.

Rice, C. E. (1997). The scientist-practitioner split and the future of psychology. *American Psychologist, 52,* 1173–1181.

Richardson, M., & Johnson, M. (1984). Counseling women. In S. Brown & R. Lent (Eds.), *Handbook of counseling psychology* (pp. 832–877). New York: Wiley.

Ridley, C. R. (2005). *Overcoming unintentional racism in counseling and therapy* (2nd ed.). Thousand Oaks, CA: Sage.

Robertiello, R., & Schoenewolf, G. (1987). *101 Common therapeutic blunders.* Northvale, NJ: Aronson.

Robitschek, C. G., & McCarthy, P. R. (1991). Prevalence of counselor self-reference in the therapeutic dyad. *Journal of Counseling and Development, 69,* 218–221.

Rodolfa, E., Hall, T., Holms, V., Davena, A., Komatz, D., & Antunez, M., et al. (1994). The management of sexual feelings in therapy. *Professional Psychology: Research and Practice, 25,* 168–172.

Rodolfa, E., Kitzrow, M., Vohra, S., & Wilson, B. (1990). Training interns to respond to sexual dilemmas. *Professional Psychology: Research and Practice, 21,* 313–315.

Rodolfa, E., Yurich, J., & Reilley, R. (1993). *Training psychologists to respond to sexual dilemmas.* Paper presented at the Annual Convention of the American Psychological Association, Toronto.

Rosenthal, D., & Frank, J. (1958). Psychotherapy and the placebo effect. In C. Reed, I. Alexander, & S. Tomkins (Eds.), *Psychopathology: A source book* (pp. 463–473). Cambridge, MA: Harvard University Press.

Rotter, J. C., & Boveja, M. E. (1999). Counseling military families. *Family Journal: Counseling and Therapy for Couples and Families, 7,* 379–382.

Sampson, J. P. (1999). Integrating Internet-based distance guidance with services provided in career centers. *Career Development Quarterly, 47,* 243–254.

Sampson, J., Kolodinsky, R. W., & Greeno, B. P. (1997). Counseling and the information highway: Future possibilities and potential problems. *Journal of Counseling and Development, 75,* 203–212.

Samstag, L. W., Batchelder, S. T., Muran, J. C., Safran, J. D., & Winston, A. (1998). Early identification of treatment failures in short-term psychotherapy: An assessment of therapeutic alliance and interpersonal behavior. *Journal of Psychotherapy Practice and Research, 7,* 126–143.

Satir, V. (1988). *New people-making* (2nd ed.). Palo Alto, CA: Science and Behavior Books.

Schmidt, L., & Meara, N. (1984). Ethical, professional, and legal issues in counseling psychology. In S. Brown & R. Lent (Eds.), *Handbook of counseling psychology* (pp. 56–96). New York: Wiley.

Schneider Corey, M., & Corey, G. (2005). *Groups: Process and practice* (7th ed.). Pacific Grove, CA: Brooks/Cole.

Schwarz, N. (1999). Self-reports: How the questions shape the answers. *American Psychologist, 54,* 93–105.

Sexton, T., & Whiston, S. (1994). The status of the counseling relationship: An empirical review, theoretical implications, and research directions. *The Counseling Psychologist, 22,* 6–78.

Sexton, T. L., Whiston, S. C., Bleuer, J. C., & Walz, G. R. (1997). *Integrating outcome research into counseling practice and training.* Alexandria, VA: American Counseling Association.

Shertzer, B., & Stone, S. (1980). *Fundamentals of counseling* (3rd ed.). Boston: Houghton Mifflin.

Shore, K. (1996). Beyond managed care and managed competition. *The Independent Practitioner: Bulletin of the Division of Independent Practice, Division 42 of the American Psychological Association, 16,* 24–25.

Silverstein, J. L. (1998). Countertransference in marital therapy for infidelity. *Journal of Sex and Marital Therapy, 24,* 293–301.

Singer, J. A., & Blagov, P. S. (2004). Self-defining memories, narrative identity, and psychotherapy: A conceptual model, empirical investigation, and case report. In L. E.

Angus & J. McLeod (Eds.), *The handbook of narrative and psychotherapy* (pp. 229–246). Thousand Oaks, CA: Sage.

Slaikeu, K. (1990). *Crisis intervention* (2nd ed.). Boston: Allyn & Bacon.

Slovenko, R. (1988). The therapist's duty to warn or protect third persons. *Journal of Psychiatry and Law, 16,* 139–209.

Smith, E. J. (2006). The strength-based counseling model. *The Counseling Psychologist, 34,* 13–79.

Smith, M., & Glass, G. (1977). Meta-analysis of psychotherapy outcome studies. *American Psychologist, 32,* 752–760.

Smyth, J. M. (1998). Written emotional expression: Effect sizes, outcome types, and moderating variables. *Journal of Consulting and Clinical Psychology, 66,* 174–184.

Sommers-Flanagan, J., & Sommers-Flanagan, R. (2002). *Clinical interviewing* (3rd ed.). New York: Wiley.

Spengler, P. M., Strohmer, D. C., Dixon, D. N., & Shivy, V. A. (1995). A scientist-practitioner model of psychological assessment: Implications for training, practice and research. *The Counseling Psychologist, 23,* 506–534.

Staats, A. (1983). *Psychology's crisis of disunity: Philosophy and method for a unified science.* New York: Praeger.

Stanard, R., & Hazler, R. (1995). Legal and ethical implications of HIV and duty to warn for counselors: When does *Tarasoff* apply? *Journal of Counseling and Development, 73,* 397–400.

Steenbarger, B. N. (1994). Duration and outcome in psychotherapy: An integrative review. *Professional Psychology: Research and Practice, 25,* 111–119.

Stevens, D. T., & Lundberg, D. J. (1998). The emergence of the Internet: Enhancing career counseling education and services. *Journal of Career Development, 24,* 195–208.

Stiles, W. B., Agnew-Davies, R., Hardy, G. E., Barkham, M., & Shapiro, D. A. (1998). Relations of the alliance with psychotherapy outcome: Findings in the second Sheffield Psychotherapy Project. *Journal of Consulting and Clinical Psychology, 66,* 791–802.

Stoltenberg, C. D., McNeill, B., & Delworth, U. (1998). *IDM supervision: An integrated developmental model for supervising counselors and therapists.* San Francisco: Jossey-Bass.

Stricker, G., & Gold, J. R. (2001). An introduction to psychotherapy integration. *NYS Psychologist, 12,* 7–12.

Strong, S. R. (1968). Counseling: An interpersonal influence process. *Journal of Counseling Psychology, 15,* 215–224.

Strunk, W., Jr., & White, E. B. (2000). *The elements of style* (4th ed.). New York: Macmillan.

Sue, D. W. (1990). Culture-specific strategies in counseling: A conceptual framework. *Professional Psychology Research and Practice, 21,* 424–433.

Sue, D. W., Ivey, A. E., & Pedersen, P. B. (Eds.). (1996). *A theory of multicultural counseling and therapy*. Pacific Grove, CA: Brooks/Cole.

Sue, D. W., & Sue, D. (2002). *Counseling the culturally diverse: Theory and practice* (4th ed.). New York: Wiley.

Sue, D., & Sundberg, N. D. (1996). Research and research hypotheses about effectiveness in intercultural counseling. In P. B. Pedersen, J. G. Draguns, W. J. Lonner, & J. E. Trimble (Eds.), *Counseling across cultures* (4th ed., pp. 323–352). Thousand Oaks, CA: Sage.

Sue, S., Zane, N., & Young, K. (1994). Research on psychotherapy with culturally diverse populations. In A. E. Bergin & S. L. Garfield (Eds.), *Handbook of psychotherapy and behavior change* (4th ed., pp. 783–820). New York: Wiley.

Sullivan, H. S. (1970). *The psychiatric interview*. New York: Norton.

Swenson, E. V., & Ragucci, R. (1984). Effects of sex-role stereotypes and androgynous alternatives on mental health judgments of psychotherapists. *Psychological Reports, 54*, 475–481.

Teyber, E. (2005). *Interpersonal process in psychotherapy: A guide for clinical training* (5th ed.). Pacific Grove, CA: Brooks/Cole.

Trierweiler, S. J., & Stricker, G. (1998). *The scientific practice of professional psychology*. New York: Plenum.

Van Kaam, A. (1966). *The air of existential counseling*. Wilkes-Barre, PA: Dimension Books.

Varenhorst, B. (1984). Peer counseling: Past promises, current status, and future directions. In S. Brown & R. Lent (Eds.), *Handbook of counseling psychology* (pp. 716–750). New York: Wiley.

Wachtel, P. L. (1993). *Therapeutic communication: Knowing what to say when*. New York: Guilford.

Wachtel, P. L. (1999). Resistance as a problem for practice and theory. *Journal of Psychotherapy Integration, 9*, 103–117.

Wachtel, P. L., & Messer, S. B. (Eds.). (1997). *Theories of psychotherapy: Origins and evolution*. Washington, DC: American Psychological Association.

Walborn, F. S. (1996). *Process variables: Four common elements of counseling and psychotherapy*. Pacific Grove, CA: Brooks/Cole.

Walitzer, K. S., Dermen, K. H., & Conners, G. J. (1999). Strategies for preparing clients for treatment: A review. *Behavior Modification, 23*, 129–151.

Walter, J. L., & Peller, J. E. (1992). *Becoming solution-focused on brief therapy*. New York: Brunner/Mazel.

Wampold, B. E., Mondin, G. W., Moody, M., Stich, F., Benson, K., & Ahn, H. (1997). A meta-analysis of outcome studies comparing bona fide psychotherapies: Empirically, "All Must Have Prizes." *Psychological Bulletin, 122*, 203–215.

Watkins, C. E., & Schneider, L. J. (1989). Self-involving versus self-disclosing counselor statements during an initial interview. *Journal of Counseling and Development, 67*, 345–349.

Watzlawick, P., Weakland, J., & Fisch, R. (1974). *Change: Principles of problem formation and problem resolution*. New York: Norton.

Welfel, E. R. (2005). *Ethics in counseling and psychotherapy: Standards, research, and emerging issues* (3rd ed.). Pacific Grove, CA: Brooks/Cole.

Wells, M. G., Burlingame, G. M., Lambert, M. J., Hoag, M. J., & Hope, C. A. (1996). Conceptualization and measurement of patient change during psychotherapy: Development of the Outcome Questionnaire and Youth Outcome Questionnaire. *Psychotherapy, 33*, 275–283.

Westefeld, J. S., Range, L. M., Rogers, J. R., Maples, M. R., Bromley, J. L., & Alcorn, J. (2000). Suicide. *The Counseling Psychologist, 28*, 445–510.

White, M., & Epston, D. (1990). *Narrative means to therapeutic ends*. New York: Norton.

Whiteley, J. (Ed.). (1982). Supervision in counseling I. *The Counseling Psychologist, 10*, 1.

Wiener, M., Budney, S., Wood, L., & Russell, R. L. (1989). Nonverbal events in psychotherapy. *Clinical Psychology Review, 9*, 487–504.

Wiger, D. (2005). *The psychotherapy documentation primer* (2nd ed.). New York: Wiley.

Wolpe, J. (1990). *The practice of behavior therapy* (4th ed.). New York: Pergamon Press.

Worden, J. W. (2008). *Grief counseling and grief therapy*. New York: Springer.

Worell, J., & Remer, P. (2002). *Feminist perspectives in therapy* (2nd ed.). New York: Wiley.

Yalom, I., & Elkin, G. (1990). *Every day gets a little closer: A twice-told therapy*. New York: Basic Books.

Yalom, I., & Leszcz, M. (2005). *The theory and practice of group psychotherapy* (5th ed.). New York: Basic Books.

Young, J. E., Klosko, J. S., Weishaar, M. E. (2003). *Schema therapy: A practitioner's guide*. New York: Guilford.

Young, J. S., & Borders, L. D. (1999). The intentional use of metaphor in counseling supervision. *Clinical Supervisor, 18*, 137–149.

Zeig, J. (Ed.). (1997). *The evolution of psychotherapy: The third conference*. New York: Brunner/Mazel.

Zetzel, E. (1956). Current concepts of transference. *International Journal of Psychoanalysis, 37*, 369–376.

INDEX